Current Topics in Microbiology and Immunology

112

Editors

Retroviruses 3

Edited by P.K. Vogt and H. Koprowski

With 19 Figures

Springer-Verlag
Berlin Heidelberg NewYork Tokyo 1984

Professor Dr. PETER K. VOGT
University of Southern California
School of Medicine
Department of Microbiology
2025 Zonal Avenue HMR 401
Los Angeles, CA 90033, USA

Professor Dr. HILARY KOPROWSKI
The Wistar Institute
36th Street at Spruce
Philadelphia, PA 19104, USA

ISBN-13: 978-3-642-69679-4 e-ISBN-13: 978-3-642-69677-0
DOI: 10.1007/978-3-642-69677-0

Library of Congress Catalog Card Number 15-12910
Softcover reprint of the hardcover 1st edition 1984

Typesetting, printing and bookbinding:
Universitätsdruckerei H. Stürtz AG, Würzburg
2123/3130-543210

Table of Contents

J. GHYSDAEL, C. BRUCK, R. KETTMANN, A. BURNY: Bovine Leukemia Virus 1

S. RUSCETTI, L. WOLFF: Spleen Focus-Forming Virus: Relationship of an Altered Envelope Gene to the Development of a Rapid Erythroleukemia. With 2 Figures 21

S.P. GOFF: The Genetics of Murine Leukemia Viruses 45

R. MÜLLER, I.M. VERMA: Expression of Cellular Oncogenes. With 17 Figures 73

Indexed in Current Contents

List of Contributors

BRUCK, C., Laboratoire de Chimie Biologique, Université Libre de Bruxelles, Brussels, Belgium

BURNY, A., Faculté des Sciences Agronomiques de l'Etat, Chembloux, Belgium

GHYSDAEL, J., Laboratoire de Chimie Biologique, Université Libre de Bruxelles, Brussels, Belgium

GOFF, S.P., Department of Biochemistry and Institute for Cancer Research, Columbia University, College of Physicians and Surgeons, 630 West 168th Street, New York, NY 10032, USA

KETTMANN, R., Faculté des Sciences Agronomiques de l'Etat, Chembloux, Belgium

MÜLLER, R., European Molecular Biology Laboratory, Postfach 102209, D-6900 Heidelberg

RUSCETTI, S., Laboratory of Genetics, National Cancer Institue, Bethesda, MD 20205, USA

VERMA, I.M., Molecular Biology and Virology Laboratory, The Salk Institute, P.O. Box 85800, San Diego, CA 92138, USA

WOLFF, L., Laboratory of Genetics, National Cancer Institute, Bethesda, MD 20205, USA

Bovine Leukemia Virus

J. GHYSDAEL[1], C. BRUCK[1], R. KETTMANN[2], and A. BURNY[2]

1	Introduction	1
2	BLV as the Etiological Agent of Enzootic Bovine Leukemia	2
2.1	BLV in Its Natural Host	2
2.2	BLV in Heterologous Hosts	3
3	Methods of Detecting BLV Infection	4
3.1	Direct Methods of BLV Assay	4
3.1.1	Syncytia Induction by BLV	4
3.1.2	Hemagglutination by BLV	5
3.2	Detection of BLV Antibodies	5
4	BLV Genome and Gene Products	6
4.1	Production of BLV in Tissue Culture	6
4.2	BLV Genome	6
4.3	BLV Proteins	7
4.3.1	Internal Structural Proteins	7
4.3.2	Reverse Transcriptase	8
4.3.3	Envelope Glycoproteins	9
4.4	Relatedness of BLV to Other Retroviruses	10
5	Molecular Aspects of BLV-Induced Leukemogenesis	10
6	Epidemiology and Transmission of BLV	12
7	Eradication and Prevention of BLV Infection	13
References		14

1 Introduction

The most common neoplasm of the bovine species is, by far, lymphoid leukosis. Two classes of bovine leukosis have been distinguished on the basis of their pathology and seroepidemiology. Sporadic bovine leukosis (SBL) is observed throughout the world as solitary cases within a herd. It is a noncontagious disease with no known viral etiology. Lymphoid tumors, multicentric, cutaneous, or thymic, are generally observed in young animals (4 months to 1 year old) and tumor cells appear to be of non-B-cell lineage. Enzootic bovine leukosis (EBL) is a highly contagious disease induced by a retrovirus exogenous to the bovine species, bovine leukemia virus (BLV). It is a chronic disease, evolving over extended periods (1–8 years), with tumors developing in only a small number of infected animals.

1 Laboratoire de Chimie Biologique, Université Libre de Bruxelles, Brussels, Belgium
2 Faculté des Sciences Agronomiques de l'Etat, Gembloux, Belgium

Current Topics in Microbiology and Immunology, Vol. 112

In this review, we will discuss what is known about enzootic bovine leukosis with special focus on the properties of its etiological agent, BLV. This knowledge has had and will have obvious applications in the development of more adequate methods for the control and eradication of EBL. Besides these practical considerations and because EBL is caused in an essentially outbred species by a virus spreading in natural conditions, this system provides an interesting model for both persistent viral infection and retrovirus-induced leukemia at the cellular, individual, and population levels.

2 BLV as the Etiological Agent of Enzootic Bovine Leukemia

2.1 BLV in Its Natural Host

In natural conditions, infection of healthy cattle by BLV occurs mainly horizontally as well as congenitally (see Sect. 6). Once infected, the animals, although showing no sign of overt viremia, elicit a strong and persistent humoral response to BLV structural proteins. The site and mechanism of BLV persistence are unknown. The majority of infected animals remain healthy and show no economic down performances (LANGSTON et al. 1978; HUBER et al. 1981; BURRIDGE et al. 1982). Depending on several ill-defined parameters, including genetic and environmental factors, between 30% and 70% of BLV carriers develop a form of the disease known as persistent lymphocytosis. Persistent lymphocytosis (PL) is characterized by an increase in circulating B-lymphocytes as evidenced by the presence of surface immunoglobulins and Fc-receptors on the circulating cells. Depending on the animal, this B-lymphocyte population represents 40–80% of the total lymphocyte population (vs 15–20% in healthy individuals) (MUSCOPLAT et al. 1974; WEILAND and STRAUB 1975; PAUL et al. 1977a, b; KENYON and PIPER 1977; KUMAR et al. 1978). BLV is present in a subpopulation of circulating B-lymphocytes of animals in PL (PAUL et al. 1977a, b; KETTMANN et al. 1978; KETTMANN et al. 1980a), where its genetic information is found integrated at a large number of sites in the cellular DNA (KETTMANN et al. 1980a).

Besides being polyclonal with respect to the integration site of the provirus, circulating PL lymphocytes exhibit no chromosomal abnormalities (HARE et al. 1967), properties that clearly contrast with those made at the tumor stage of the disease (see below).

Only a few percent (0.1–10%, depending on nondefined genetic factors) of BLV-infected animals, whether or not in PL, develop lymphoid tumors. Extensive studies of the symptoms and the anatomopathological and histopathological lesions of BLV-induced tumors have been reviewed previously (ANDERSON et al. 1968; FUJIMOTO et al. 1969; JARRETT and MACKEY 1974; WITTMAN and URBANECK 1969; MAMMERICKX 1973; BURNY et al. 1978a).

Based on cytological, cytoenzymological, and immunochemical methods, and by analogy to the Kiel classification of human malignant lymphomas (GERARD-MARCHAND et al. 1974), lymphocytes of BLV-induced tumors appear to be of the B-cell lineage (PARODI et al. 1982), with a nonuniform phenotype of differentiation for tumor cells of different animals (TAKASHIMA et al. 1977; ONUMA et al. 1979; PARODI et al. 1982). Southern blot analysis of bovine tumor DNA for BLV proviral sequences demonstrates the viral genetic information to be present in most of the cells of all tumors (KETTMANN et al. 1980a). Unlike the situation encountered in PL lymphocytes, tumor cells are mono- or oligoclonal for the site of BLV integration (KETTMANN et al. 1980a). Individual tumors may contain one to several provirus(es), either complete or harboring deletions and, when they occur, deletions affect the 5′ half of the proviral genetic information (KETTMANN et al. 1982). No evidence has been observed so far for a common integration site for BLV provirus in different tumors (see Sect. 5). Defects in the ploidy of the chromosomal patterns are frequently detected but do not appear to follow a common pattern in different tumors (HARE et al. 1967).

In both PL and tumor cells, BLV proviral information appears to be repressed (DRISCOLL and OLSON 1977; BALIGA and FERRER 1977; DEVARE et al. 1978), and this lack of expression is controlled at the transcriptional level (KETTMANN et al. 1980b; MARBAIX et al. 1981; KETTMANN et al. 1982). When isolated from their host and maintained in short-term tissue culture, PL lymphocytes and in some cases tumor cells (C. Bruck, unpublished information) express viral mRNA and proteins and at least some of them produce viral particles (MILLER et al. 1969; STOCK and FERRER 1972; DRISCOLL and OLSON 1977). This lack of expression clearly prevents cell lysis by antibody-mediated cytolytic mechanisms and allows the expansion of BLV-carrying cell clones. Indeed, all infected animals have high titer antibodies to BLV structural proteins (see Sect. 3), and antibodies to the major envelope glycoprotein of the virus exhibit a strong cytolytic activity to BLV-producing cells, including tumorous lymphocytes in tissue culture (PORTETELLE et al. 1978, 1980; C. Bruck, unpublished results). The mechanism by which BLV proviral transcription is repressed in PL and tumor lymphocytes is not understood. Besides the clear interference by virus-neutralizing antibodies, recent experiments indicate that plasma of BLV-infected cattle specifically contains a nonimmunoglobulin protein molecule that blocks the transcription of the viral genome in cultured lymphocytes (GUPTA and FERRER 1982; J. Ferrer, personal communication). The continuous antigenic stimulation of the immune system of infected animals implies antigen production by cell types either localized in areas inaccessible to antibody molecules or resistant to antibody-mediated lysis. No experimental data exist about the nature of these antigens (soluble or in viral particles) or the site of their production.

2.2 BLV in Heterologous Hosts

Natural cases of malignant lymphomas in sheep are rare and the viruses isolated from these cases appear indistinguishable from BLV (reviewed by BURNY et al. 1980).

In experimental conditions, however, sheep can be easily infected by BLV, usually with increasing titers of antibodies to viral structural proteins from the time of infection onward (reviewed by Burny et al. 1980). A major difference exists between sheep and cattle pathology in that PL is a situation never observed in sheep. Also, tumors arise at a higher frequency and after a much shorter latent period (1–3 years) than observed in cattle. Anatomopathological lesions are comparable for tumors of sheep and cattle, but recent evidence seems to indicate that, unlike BLV-induced tumor cells in cattle, tumor cells in sheep are of T-cell origin (Horvath 1982). As in the bovine species, BLV-induced tumor cells in sheep are mono- or oligoclonal for BLV provirus integration site, and the proviral genetic information is not transcribed (R. Kettmann, unpublished results). Only one case of lymphosarcoma development has been reported in goats experimentally infected with BLV. Lymphoid tumors appeared 8 years postinfection in various organs and all tumorous tissues contained the same BLV-containing clone (Olson et al. 1981).

Experimental infections with BLV as evidenced by persistent antibody production has been reported for chimpanzees (van der Maaten and Miller 1976a), pigs (Mammerickx et al. 1981; Baumgartener and Olson 1982), domestic rabbits, cats, dogs, deer, rats, and guinea pigs (Baumgartener and Olson 1982). In none of these species were overt viremia and pathological disorders observed. No evidence for the production of BLV antibodies following BLV inoculation was reported for mouse, chipmunk, ground squirrel, Japanese quail, and chicken (Baumgartener and Olson et al. 1982).

Infection by BLV has never been reported for human beings, even those potentially exposed to infectious virus or virus-producing material (reviewed by van der Maaten and Miller 1977).

3 Methods of Detecting BLV Infection

3.1 Direct Methods of BLV Assay

3.1.1 Syncytia Induction by BLV

Syncytia induction (polykaryocytosis) by BLV, originally observed in tissue culture established from BLV-induced tumors (Cornefert-Jensen et al. 1969; van der Maaten et al. 1974; Ressang et al. 1974), has been studied in detail in several laboratories. Depending on the assay conditions, two types of phenomena have been described as early and late polykaryocytosis.

Late polykaryocytosis is induced on appropriate monolayers of indicator cells either by BLV-infected lymphocytes or cell-free preparations of BLV; syncytia usually appear 4–8 days after initiation of the cocultures or postinfection, respectively (Diglio and Ferrer 1976). This method has been used as a BLV infectivity assay, and since syncytia formation is specifically inhibited by natural antibodies to BLV it has also been used for the detection of BLV antibodies in infected animals (Benton et al. 1978; Ferrer et al. 1977). Early polykaryocytosis is induced in appropriate mammalian sarcoma virus-transformed cells cocultivated with BLV-producing cells (Guillemain et al. 1977; Irgens et al. 1977;

OGURA et al. 1977) or infected with cell-free BLV preparations (GRAVES and JONES 1981). Syncytia appear a few hours postinoculation, are usually very large, and develop in a dose-dependent function of BLV-infected effector cells (GUILLEMAIN et al. 1977). As for late polykaryocytosis, early polykaryocytosis is inhibited by sera containing antibodies to BLV gp51, and recent experiments have shown that monoclonal antibodies to BLV gp51 that neutralize infectivity also inhibit early polykaryocytosis (BRUCK et al. 1982b; see Sect. 3.2).

3.1.2 Hemagglutination by BLV

Recent experiments show that cell-free preparations of BLV are able to agglutinate mouse erythrocytes specifically (SENTSUI et al. 1982). The BLV hemagglutinin appears to be BLV gp51 since purified gp51 has hemagglutinating activity.

3.2 Detection of BLV Antibodies

Antibodies to virtually all known BLV structural proteins are present in the sera of infected cattle and sheep (reviewed by BURNY et al. 1978a, 1980). In most cases, antibody titers are the highest to the gp51 major envelope glycoprotein. Several lines of evidence indicate that the natural antibody response of infected animal is directed toward a very limited number of possible antigenic regions of gp51:

1. Natural antibodies to gp51 do not compete for the same antigenic sites on gp51 as antibodies prepared in rabbits to purified gp51 (PORTETELLE et al. 1980).
2. Among mouse monoclonal antibodies prepared against purified gp51 and recognizing eight distinct antigenic sites on the molecule (BRUCK et al. 1982a; see Sect. 4), only those recognizing epitopes F, G, and H were shown to be able to compete with sera of BLV-infected cattle for gp51 binding (C. Bruck, unpublished data). Monoclonal antibodies against the five other gp51 epitopes do not appear to have functional homologues in the sera of BLV-infected cattle.

Epitopes F, G, and H are precisely those recognized by neutralizing antibodies to both virus infectivity and syncytia formation and, at least one of them, epitope H, is also a target for cytolytic antibodies (BRUCK et al. 1982b).

Since persistent antibodies to BLV proteins are a constant and early feature of BLV infection in experimental and natural conditions, serological surveys of cattle populations are the best method for early detection of field BLV infection. Early detection of BLV infection is important for both the control of spread of the disease and the establishment of an efficient eradication policy.

Numerous methods have been developed for the detection of antibodies specific to BLV in the sera of BLV-infected cattle or sheep, including agar gel immunodiffusion, liquid-phase or solid-phase radioimmunoassays, and enzyme-linked immunosorbent assays (reviewed by Burny et al. 1980). A recent variant to classical enzyme-linked immunosorbent assays has been developed recently that yields a highly sensitive, practical, rapid, and cheap method for the detection of BLV gp51 antibodies (Portetelle et al. 1983).

4 BLV Genome and Gene Products

4.1 Production of BLV in Tissue Culture

Since the original discovery of BLV in the supernatant of short-term cultures of PL lymphocytes (Miller et al. 1969), several long-term cultures producing BLV have been obtained. These include cell lines derived from explanted tumor tissues (Ferrer et al. 1971; Mamoun et al. 1981) or various epithelioid and fibroblastoid cell lines infected with BLV after coculture with BLV-infected bovine lymphoid cells (van der Maaten et al. 1974; van der Maaten et al. 1976b; Graves and Ferrer 1976). Most of the data about the structure and biosynthesis of BLV components have been obtained from the Fetal Lamb Kidney-BLV-producing cell line of van der Maaten and Miller (1976b). This cell line is the best available producer of infectious BLV particles. Restriction enzyme analysis indicates that it contains four BLV proviruses (Casey et al. 1982). Each of these proviruses has been molecularly cloned and only one of them proved infectious in subsequent transfection experiments.

4.2 BLV Genome

Denaturation of the 60–70S BLV virion RNA yielded poly(A)-containing subunits that sedimented at 38S and migrated as single components of molecular weight 2.95×10^6 (ca. 8.8 kb) in sodium dodecyl sulfate (SDS) polyacrylamide gels (Ghysdael et al. 1979). This value is in good agreement with the 8.3- to 9.4-kb values reported for unintegrated BLV DNA from various sources (Kettmann et al. 1981; Kashmiri et al. 1983) and for BLV proviruses cloned either from a bovine tumor (Deschamps et al. 1981) or the FLK-BLV producer cell line (Casey et al. 1982).

From the nucleotide sequence of the 3′ long terminal repeat (LTR) and portion of the 5′ LTR of a BLV provirus cloned from a bovine tumor it appears that (Couez et al. 1983):

1. In agreement with the reported size of BLV strong stop cDNA (Seiki et al. 1982), the RU_5 moiety of the viral RNA is 320 nucleotides long, thus leaving about 216 nucleotides for the U_3 region of the RNA.
2. Sequences possibly involved in the regulation of the promoter function of proviral LTR are localized 38 nucleotides ("TATA" box) and 92 nucleotides ("CAT" box) upstream from the mRNA "cap" site (first nucleotide of R).

3. A stretch of 18 nucleotides immediately downstream from U_5 is identical to the $tRNA^{Pro}$-binding site of other mammalian retroviruses, suggesting that $tRNA^{Pro}$ is used as primer for the synthesis of (−) DNA strand by reverse transcriptase.
4. A stretch of nine purine residues is present immediately upstream from U_3 that may represent a binding site for the primer of (+) DNA strand synthesis by reverse transcriptase.

The coding capacity of the viral genome includes at least the four internal structural proteins (p24, p15, p12, p10), reverse transcriptase, and two envelope glycoproteins (gp51 and gp30). Based on functional (GHYSDAEL et al. 1978, 1979) and structural (R. Gilden, personal communication) analyses, it appears that the gene order along the BLV genome is *5′-gag-pol-env-3′*, similar to that of other replication-competent lymphoid leukemia viruses. It is not clear at present whether additional proteins are encoded by the viral genome. Along these lines it should be noted that in vitro translation of 3′-derived 16–18S subgenomic fragments of the BLV 38S RNA yielded an 18000-dalton protein unrelated by peptide analysis to the products of the *gag, pol,* and *env* genes (GHYSDAEL et al. 1979). It remains to be seen, however, whether this product is synthesized in infected cells or is merely an in vitro translation artefact.

4.3 BLV Proteins

Analysis of purified BLV virions by SDS polyacrylamide gel electrophoresis and gel filtration in the presence of guanidine hydrochloride has resolved four major nonglycosylated proteins (p24, p15, p12, p10) and two glycosylated components (gp51 and gp30) (reviewed by BURNY et al. 1980).

4.3.1 Internal Structural Proteins

p24 was the first recognized BLV structural protein (MILLER and OLSON 1972). It has been purified to homogeneity and extensively characterized (GILDEN et al. 1975; DEVARE et al. 1976; MCDONALD and FERRER 1976; PORTETELLE et al. 1977). It is a neutral and moderately hydrophobic protein and has been partially sequenced (OROSZLAN et al. 1979). Although BLV p24 has been shown to share the aminoterminal proline and carboxyterminal leucine of the p30 of all mammalian retroviruses, in agreement with previous immunochemical analyses, it has been found to be chemically distinct from the major core protein of other retroviruses. However, statistically significant sequence homology was found between BLV p24 and the major p24 core protein of human T-cell leukemia virus (HTLV) (OROSZLAN et al. 1982). Monoclonal antibodies to BLV p24 have been recently obtained that identify two distinct antigenic sites on purified p24 (Bruck et al., unpublished results). Antibodies to these epitopes are present in the sera from BLV-infected animals.

p15 is a slightly basic protein (KAADEN et al. 1977) that lacks methionine residues. It is the major phosphoprotein of BLV virions (UCKERT and WUNDERLICH 1979) and chemical cross-linking experiments indicate that two populations of p15 molecules exist, one linked to the lipid bilayer of the viral membrane and the other linked to viral RNA (UCKERT et al. 1983).

p12 is a basic protein with a high content of glycine and proline residues and thus a probable low level of secondary structure. Chemical cross-linking experiments have shown p12 to be in close proximity to viral RNA in BLV particles (UCKERT et al. 1983). Comparison of the p12 protein sequence with that derived from the partial nucleotide sequence of the *gag* gene of a cloned tumor provirus has localized p12 at the carboxy terminus of the primary translation product of *gag* (R. Gilden, personal communication). These experiments also point to minimal variations (one amino acid difference) between the p12 sequences of the two BLV isolates. This minimal genetic drift among independent isolates further supports the significance of the homology detected between BLV p12 and HTLV p12 (OROSZLAN et al. 1983; see Sect. 4.4).

p10 appears to be a highly basic protein able to bind to single-stranded DNA (LONG et al. 1980).

Intracellular precursors to BLV virion proteins have been detected by immunoprecipitation with sera of defined specificity of extracts of both BLV-infected cells (GHYSDAEL et al. 1978; MAMOUN et al. 1983) and *Xenopus laevis* oocytes microinjected with BLV 38S RNA (GHYSDAEL et al. 1979). In all cases, two polypeptides of molecular weight 70000 (Pr 70^{gag}) and 45000 (Pr 45^{gag}) were found to be precursors to p24, p15, p12, and p10. The relationship between Pr 70^{gag} and Pr 45^{gag} is unclear since, although the methionine-containing peptides of Pr 45^{gag} were all contained in Pr 70^{gag}, it appears unlikely that Pr 45^{gag} derived from Pr 70^{gag} by proteolytic cleavage since (a) the Pr 70^{gag}/Pr 45^{gag} ratio in infected cells was found to be the same, irrespective of the pulse-labeling time used, (b) the Pr 70^{gag}/Pr 45^{gag} ratio was unaffected by labeling the cells in the presence of aminoacid analogs known to inhibit further proteolytic cleavages, and (c) both Pr 70^{gag} and Pr 45^{gag} are synthesized in reticulocyte lysate programmed with BLV 38S RNA at a ratio similar to that found in BLV-infected cells (GHYSDAEL et al. 1978, 1979 and unpublished results).

Available evidence appears thus to indicate that Pr 70^{gag} and Pr 45^{gag} are synthesized independently, possibly on differently spliced 38S mRNA populations.

4.3.2 Reverse Transcriptase

Size estimates for the RNA-dependent DNA-polymerizing activity of BLV virions by glycerol gradient sedimentation have varied from 70000 to 45000, depending on the salt concentration used (BURNY et al. 1978b; DRESCHER et al. 1979). The RNA-dependent DNA-polymerizing activity was found strictly dependent on RNA and all four deoxyribonucleoside triphosphates and, irrespective of the template-primer used, displayed a strong preference for Mg^{++} over Mn^{++} (GILDEN et al. 1975; KETTMANN et al. 1976; GRAVES et al. 1977).

Evidence for the existence of RNaseH or endonuclease activities associated with BLV reverse transcriptase has not been investigated. A possible precursor to BLV reverse transcriptase is a 145000-dalton product, detected in BLV-infected cells and in heterologous protein-synthesizing systems programmed with BLV 38S as an elongation product of Pr 70^{gag} (GHYSDAEL et al. 1978, 1979).

4.3.3 Envelope Glycoproteins

The major envelope glycoproteins of BLV particles, gp51 and gp30, have been shown to be associated as a 1:1 complex by disulfide bonds in viral particles isolated under nonreducing conditions (reviewed by BURNY et al. 1978).

Immunoprecipitation experiments using a monospecific antiserum to BLV envelope glycoproteins indicate that both gp30 and pg51 derive from a glycosylated 72000-dalton precursor gPr 72^{env} (GHYSDAEL et al. 1978; MAMOUN et al. 1983).

Treatment of FLK-BLV cells with tunicamycin, an inhibitor of the first step in the pathway of the synthesis of the oligosaccharide moiety of asparagine-linked glycoprotein, has made possible the determination of the size of the apoprotein of gPr 72^{env} as a 48000-dalton component (BRUCK et al. 1984).

Monoclonal antibodies to BLV gp51 have been obtained in the mouse using either nondisrupted BLV particles or purified gp51 as immunogens (BRUCK et al. 1982a).

By competition antibody-binding assays, these antibodies were shown to identify eight independent antigenic sites (epitopes) on the gp51 molecule (sites A through H, BRUCK et al. 1982a). Three of these eight antigenic sites, namely F, G, and H, were shown to be involved in determining both virus infectivity and syncytia formation by BLV-producing cells, whereas at least site H appeared to be a target for antibody-mediated complement-dependent cytotoxicity (BRUCK et al. 1982b).

Limited proteolytic digestion of gp51 with urokinase and immunoprecipitation studies with the different monoclonal antibodies have shown that antigenic sites F, G, and H, together with site E, are part of a 15000-dalton hypoglycosylated fragment of gp51 (BRUCK et al. 1982b).

However, integrity of the F, G, and H antigenic sites appears to be determined by the integrity of a carbohydrate moiety since (a) removal of most of the carbohydrate chains of gp51 by treatment with a mixture of glycosidases specifically abolished the reactivity of the protein with sera of infected cattle and sheep (PORTETELLE et al. 1980; SCHMERR et al. 1981) and (b) the nonglycosylated 48000-dalton protein moiety of the precursor to gp51 synthesized after tunicamycin treatment of FLK-BLV cells, although recognized by monoclonal antibodies to antigenic sites A–E, is not immunoprecipitable by monoclonal antibodies to epitopes F, G, and H (BRUCK et al. 1984).

4.4 Relatedness of BLV to Other Retroviruses

Nucleic acid hybridization experiments as well as competition radioimmunoassays have failed to detect any homology between BLV and feline leukemia virus, the Rauscher strain of murine leukemia virus, mouse mammary tumor virus, avian myeloblastosis virus (standard strain), baboon endogenous virus, woolly monkey leukemia virus, and Mason-Pfizer monkey virus, as well as murine and porcine endogenous viruses (KAWAKAMI et al. 1970; KETTMANN et al. 1976; MCDONALD and FERRER 1976; DEVARE et al. 1977).

However, as mentioned in the preceding section partial sequence analysis of BLV p24 has revealed statistically significant but distant homology to the major core protein of other mammalian viruses (OROSZLAN et al. 1979). More extensive homology has been reported to the major p24 core protein of HTLV-1, a human retrovirus associated with adult T-cell lymphomas (POIESZ et al. 1980, 1981; HINUMA et al. 1981).

Even more important homology has been found between the amino acid sequences of BLV and HTLV-1 p12s (OROSZLAN et al. 1983), detectable in competition radioimmunoassays (OROSZLAN et al. 1982).

Whether the evolutionary relationship between the *gag* genes of BLV and HTLV can be extended to other viral genes is not known yet. The complete nucleotide sequence of the BLV LTR has been obtained recently (COUEZ et al. 1984), and no homology was found to the sequences of the LTRs of other retroviruses, including HTLV-1.

By various techniques of molecular hybridization, BLV genetic information has been found to be totally exogenous to the genome of all vertebrate species tested (KETTMANN et al. 1976, 1978; CALLAHAN et al. 1976; DESCHAMPS et al. 1981). Completion of the total nucleotide sequence of the BLV genome should make possible the search for discrete regions of homology to cellular DNA sequences and possibly shed light on the evolutionary origin of the virus.

5 Molecular Aspects of BLV-Induced Leukemogenesis

This section summarizes the experiments made toward the understanding of BLV-induced leukemia at the molecular level.

As mentioned in the previous section, the use of probes specific to the BLV genome in hybridization analyses of cellular DNAs has provided evidence for the exogenous character of BLV genetic information to the DNA of all vertebrate species tested. This conclusion has been definitively established by experiments using as probe a molecularly cloned BLV provirus isolated from a BLV-induced tumor harboring a single provirus (DESCHAMPS et al. 1981). These results are of importance to the possible origin of BLV but also demonstrate that BLV genetic information includes no *onc* genes since all retroviral *onc* genes are cell-derived, phylogenetically conserved, transformation-specific inserts in the genome of acute transforming viruses (reviewed by COFFIN 1981).

Liquid hybridization and Southern blots analyses using BLV-genome-specific probes show that all BLV-induced tumors and most of the cells in individual tumors contain proviral genetic information, thus demonstrating the indispensable presence of the virus for tumor development (KETTMANN et al. 1980, 1981). Furthermore, individual tumors are clonal for the site of provirus integration, a situation in sharp contrast to the polyclonality observed in lymphocytes of either BLV-infected animals or animals in PL (KETTMANN et al. 1980). Also, unlike BLV-infected lymphocytes of animals in PL, tumor lymphocytes display abnormalities in their chromosomal ploidy, although no constant pattern can be detected (HARE et al. 1967).

These observations point to a multistep process for tumor development, the presence of the BLV provirus being a necessary but, by itself, an insufficient event.

Recent experiments have evaluated the role of BLV proviral information in the development and/or maintenance of the tumor state.

BLV-induced tumors contain at least one or a portion of one provirus and, when they occur, deletions seem to affect the 5′ half (but possibly not the 5′-LTR) of the provirus (KETTMANN et al. 1980). Whether the apparent generalized conservation of 3′ proviral sequences is of functional importance at a particular stage of disease development is at present unknown. Irrespective of the method used (liquid or dot blot hybridization of poly $(A)^+$ RNA from independent tumors), transcription of BLV proviral DNA, including LTR sequences, was not detected (< one copy/cell; KETTMANN et al. 1981, 1982; MARBAIX et al. 1981). This indicates that maintenance of the tumor state does not require (a) continuous expression of BLV genetic information and (b) continuous LTR-mediated promotion of transcription into cellular sequences 3′ to the provirus, a mechanism known to be operative in the activation of c-*myc* transcription in most avian leukosis virus (LLV)-induced avian lymphomas (HAYWARD et al. 1981).

It should be noted that these experiments have not addressed the question as to the role in tumors of LTR-mediated enhancement of transcription from cellular promoters either 3′ or 5′ to the provirus or of LTR-mediated mRNA stabilization (e.g., by polyadenylation), mechanisms that are probably also involved in c-*myc* activation in some LLV-induced avian lymphomas (PAYNE et al. 1981).

Also, the possibility that integration of the BLV provirus results in an inhibitory effect on the transcription of nearby cellular genes remains untested. If any of these hypothetical mechanisms is operative in the deregulation of transcription of cellular genes important to the control of growth and/or differentiation, all imply as corollary the integration of at least one BLV provirus in a limited number of domains in the cellular DNA of independent tumors.

The existence of preferential integration sites for BLV proviruses in tumor-cell DNAs has been investigated (KETTMANN et al. 1982, 1983). The 3′-end host-virus junction fragments of two independent BLV-induced tumors, each containing a single provirus, have been molecularly cloned. The cell-derived moieties of these clones were used as probes to define large restriction fragments in normal cellular DNA (15 and 17 kb, respectively) corresponding to the cellular domains surrounding these proviruses in the original tumors. No rearrangement of these two cellular loci, due to the insertion of a BLV provirus, has been found in 28 other BLV-induced tumors. Furthermore, analysis of hamster-bovine cell hybrids (Huertz and Hors-Cayla 1981) indicate that the two cellular sequences used as probes in these experiments belong to different chromosomes in normal bovine cells (D. Gregoire et al., unpublished results).

In conclusion, experimental results have failed so far to demonstrate a preferential integration site for BLV proviruses in tumors. It appears that, if preferential integration occurs in tumors, several sites must be involved and the integration domain must be very large. Similar analysis using additional host-virus junction fragments will be necessary to reinforce this point. To summarize, these results, although confirming definitively the role of BLV in tumor induction, have not pointed yet to the function of the BLV provirus or proviral information in this process.

Along these lines, *Eco*RI analysis of a panel of ten tumor DNAs has failed to demonstrate rearrangements of the cellular homologues of several retroviral *onc* genes including c-*myc*, c-*myb*, c-*erb*A, c-*erb*B, c-*abl*, c-*src*, c-*ras*^{H}, c-*fes*, and c-*sis* (Kettmann et al. 1983). Also, transfection of high molecular weight DNA of most BLV-induced tumors into NIH 3T3 cells fails to induce foci of transformed cells (J. Ghysdael and C.J. Marshall, unpublished results).

It should be mentioned, however, that all reported experiments involve only the tumor phase of the disease of cattle. As discussed in Sect. 2, in experimental conditions BLV is highly oncogenic for sheep, where it appears to induce a T-cell lymphoma. Furthermore, as tumors in general and leukemias in particular appear to be multistep processes, it may well be that the mandatory functional role of BLV in lymphoma development occurs at a stage preceding the clonal expansion of tumor cells. These questions are clearly to be approached in future experiments.

6 Epidemiology and Transmission of BLV

As extensively reviewed previously (Burny et al. 1980), the use of highly sensitive serological tests for the detection of BLV infection has demonstrated the virus to be distributed worldwide although striking zones of prevalence are evident (e.g., eastern Europe, North and South America). Little information is available about BLV distribution in African and, except Japan, Asian countries.

Generally speaking, transmission of retroviruses to susceptible animal species can occur horizontally, congenitally, or genetically. Since BLV genetic information is totally exogenous to its natural host, vertical genetic transmission can be excluded as a mode of propagation of this virus. Vertical congenital transmission of BLV from infected dams to their progeny occurs occasionally in natural conditions; congenital transmission from infected sires to progeny has never been reported (reviewed by Burny et al. 1980).

Horizontal transmission is undoubtedly the most important form of BLV propagation both within a herd or between herds as the consequence of trade exchanges.

Several observations (HOUSE and HOUSE 1977; MAAS-INDERWIESEN et al. 1978; WILESMITH et al. 1978) point to management and husbandry practices as important factors in the rate of dissemination of infection within a herd. Natural BLV transmission can also occur through milk and colostrum from infected animals since experimental transmission to sheep is easily achieved using these secretions as inocula (MILLER and VAN DER MAATEN 1978b; STRAUB 1982; KENYON et al. 1982). Colostrum and milk transmission are very efficient if these secretions are contaminated by blood (MAMMERICKX 1972a, b). In relation to the potential hazard of BLV to human beings, it should be noted that pasteurization of infected milk totally destroys infectious viral particles (BAUMGARTENER et al. 1976).

Several species of blood-sucking insects could act as BLV vectors (BECH-NIELSEN et al. 1978; KAADEN et al. 1982). Although this mode of propagation is probably of minor importance in moderate climates, it could be more important in tropical conditions, where nonbovine species such as zebu, water buffaloes, and capybaras appear as possible BLV reservoirs (MARIN et al. 1982).

7 Eradication and Prevention of BLV Infection

Control of the spread of BLV infection is desirable for two main reasons. The first and obvious reason is economic. Although in most cases BLV infrequently induces tumors and death of infected animals, in some situations and for reasons that are essentially unclear, high tumor rates have been observed (YOSHIKAWA et al. 1972; MARÍN et al. 1978). Especially in these situations but also since the extent of tumor development is directly linked to the extent of virus spread, uncontrolled infection can lead to important economic losses.

The second reason is the potential hazard of BLV to human health. It should be stressed here that no evidence for a risk to human health associated to BLV has been obtained so far (reviewed by VAN DER MAATEN and MILLER 1977). However, evidence exists for BLV transmission to heterologous species and genera in natural conditions (MARÍN et al. 1980). Also, sequence homology exists between the genetic information of BLV and that of the only characterized human retrovirus, HTLV, making, at least theoretically, genetic recombination events between the two viruses possible.

These reasons have justified the search for eradication methods of BLV infection. Countries of the European Economic Community have declared enzootic bovine leukosis as a legally notifiable disease. In those countries where moderate to high rates of BLV infection are observed, successful eradication programs have been established. In such programs, all animals of a herd are serially tested at intervals of 3–6 months for antibodies to BLV gp51. Seropositive animals are eliminated once detected and the whole herd is retested until definitive disappearance of reactors (MAMMERICKX et al. 1978; STRAUB 1978; BURKI 1982; DE VRIES 1982; SCHMIDT 1982). For countries where BLV infection is endemic (American countries, eastern Europe countries), although similar tests involving the isolation of calves identified as BLV-free are possible (FERRER

1982), they pose management problems and will probably prove difficult to achieve in practice. In these countries in particular but in all countries in general the ideal solution toward control of BLV infection and elimination of enzootic bovine leukosis would be the use of an effective BLV vaccine.

Since antibodies to BLV gp51 are neutralizing in vitro to both virus infectivity and release and include subclasses with cytolytic activity (see Sect. 4), a vaccine resulting in the production of antibodies to gp51 should be protective to virus infection and spread.

Preliminary experiments using as vaccine chemically inactivated BLV (MILLER and VAN DER MAATEN 1978a; MILLER et al. 1983), fixed FLK cells, or purified gp51 (ONUMA et al. 1983) clearly indicate that humoral antibody response to gp51 is effective in short-term protection to BLV infection. Three antigenic sites that are part of the 15000-dalton fragment of the gp51 molecule (sites F, G, H, see Sect. 4) have been shown to be involved in virus infectivity (BRUCK et al. 1982b). Since at least one of these sites appears to be conserved in all BLV isolates tested so far (C. Bruck, unpublished data), these sites might serve as the basis of a synthetic BLV vaccine. Obvious possibilities are the insertion of the region of the BLV *env* gene corresponding to these sites in suitable plasmid expression vectors, including yeast expression vectors since glycosylation appears to be important to the biological activity of sites F–H (see Sect. 4). Alternatively, from the nucleotide sequence of the BLV *env* gene, synthetic peptides corresponding to various portions of the amino acid sequence of gp51, including antigenic sites F–H, can be tested as potential vaccine after coupling to an adequate protein carrier.

Of important practical value, the use of gp51-based vaccines would make very simple the distinction in field conditions between vaccinated animals and BLV-infected animals, the latter synthesizing antibodies not only to gp51 but also to other BLV structural proteins.

An alternative to gp51 vaccination has been proposed recently (THEILEN et al. 1982). It makes use of vaccination of cattle with live cells from a cell line, BL-3, established from the bone marrow and thymus of a sporadic case of bovine leukosis. Although injected animals had no antibodies to BLV structural proteins, protection against virus infection was evident. The nature of the antigen(s) of BL-3 conferring protection to vaccinated animals has been proposed to be tumor-associated transplantation antigen(s) (THEILEN et al. 1982).

In conclusion, preliminary experimental evidence has pointed to several avenues for the development of an effective BLV vaccine, making the production of such a vaccine undoubtedly forseeable in the next decade. Its use should achieve worldwide eradication of BLV infection and enzootic bovine leukosis.

References

Anderson LJ, Jarrett WFH, Crighton GW (1968) A classification of lymphoid neoplasms of domestic animals. Natl Cancer Inst Monogr 32:343–353

Baliga V, Ferrer J (1977) Expression of the bovine leukemia virus and its internal antigen in blood lymphocytes. Proc Soc Exp Biol Med 156:388–391

Baumgartener LE, Olson C (1982) Host range of bovine leukosis virus: preliminary report. Curr Top Vet Med Anim Sci 15:338–347

Baumgartener L, Olson C, Onuma M (1976) Effect of pasteurization and heat treatment on bovine leukemia virus. J Am Vet Med Assoc 169:1189–1191

Bech-Nielsen S, Piper CE, Ferrer JF (1978) Natural mode of transmission of the bovine leukemia virus: role of blood-sucking insects. J Am Vet Med Assoc 39:1089–1092

Benton CV, Soria AE, Gilden RV (1978) Direct syncytial assay for the quantitation of bovine leukemia virus. Infect Immun 20:307–309

Bruck C, Mathot S, Portetelle D, Berte C, Franssen JD, Herion P, Burny A (1982a) Monoclonal antibodies define eight independent antigenic regions on the bovine leukemia virus (BLV) envelope glycoprotein gp51. Virology 122:342–352

Bruck C, Portetelle D, Burny A, Zavada J (1982b) Topographical analysis by monoclonal antibodies of BLV gp51 epitopes involved in viral functions. Virology 122:353–362

Bruck C, Rensonnet N, Portetelle D, Cleuter Y, Mammerickx M, Burny A, Mamoun R, Guillemain B, Van der Maaten M, Ghysdael J (1984) Biologically active epitopes of Bovine leukemia virus glycoprotein gp51: their dependence on protein glycosylation and genetic variability. Virology, in press

Burki F (1982) Experiences and progress achieved with BLV elimination from Austrian livestock. Curr Top Vet Med Anim Sci 15:516–527

Burny A, Bex F, Chantrenne H, Cleuter Y, Dekegel D, Ghysdael J, Kettmann R, Leclercq M, Leunen J, Mammerickx M, Portetelle D (1978a) Bovine leukemia virus involvement in enzootic bovine leukosis. Adv Cancer Res 28:251–311

Burny A, Bex F, Bruck C, Cleuter Y, Dekegel D, Ghysdael J, Kettmann R, Leclercq M, Mammerickx M, Portetelle D (1978b) Biochemical studies on enzootic and sporadic types of bovine leukosis. In: Chandra P (ed) Antiviral mechanisms in the control of neoplasia. Plenum, New York, pp 83–99

Burny A, Bruck C, Chantrenne H, Cleuter Y, Dekegel D, Ghysdael J, Kettmann R, Leclercq M, Leunen J, Mammerickx M, Portetelle D (1980) Bovine leukemia virus: molecular biology and epidemiology. In: Klein G (ed) Viral oncology. Raven, New York, pp 231–289

Burridge MJ, Thurmond MC, Puhr DM, Wilcox CJ, Simerl NA (1982) Preliminary studies on impact of bovine leukemia virus infection on dairy productivity. Curr Top Vet Med Anim Sci 15:599–603

Callahan R, Lieber MM, Todaro GJ, Graves DC, Ferrer JF (1976) Bovine leukemia virus genes in the DNA of leukemic cattle. Science 192:1005–1007

Casey JW, Nicolson MO (1982) Susceptibility to productive infection by the bovine leukemia virus is governed at the proviral level. Fed Proc 41 (3):1275

Coffin JM, Varmus HE, Bishop JM, Essex M, Hardy WD, Martin GS, Rosenberg NE, Scolnick EM, Weinberg RA, Vogt PK (1981) Proposal for naming host cell-derived inserts in retrovirus genomes. J Virol 40:953–957

Copeland TD, Morgan MA, Oroszlan S (1983) Submitted for publication

Cornefert-Jensen F, Hare WC, Stock ND (1969) Studies on bovine lymphosarcoma. Formation of syncytia and detection of virus particles in mixed cell cultures. Int J Cancer 4:507–519

Couez D, Deschamps J, Kettmann R, Stephens RM, Gilden RW, Burny A (1984) Nucleotide sequence analysis of the long terminal repeat of integrated bovine leukemia virus DNA and adjacent viral and host sequences. J Virol 49:615–620

Deschamps J, Kettmann R, Burny A (1981) Experiments with cloned complete tumor-derived bovine leukemia virus information prove that the virus is totally exogenous to its target animal species. J Virol 40:605–609

Devare SG, Chander S, Samagh BS, Stephenson JR (1977) Evaluation of radioimmunoprecipitation for the detection of bovine leukemia virus infection in domestic cattle. J Immunol 119:277–282

Devare SG, Oroszlan S, Stephenson JR (1978) Application of radioimmunologic techniques to epidemiologic studies of lymphosarcoma in domestic cattle. Ann Rech Vet 9:689–697

Devare SG, Stephenson JR, Sarma PS, Aaronson SA, Chander S (1976) Bovine lymphosarcoma: development of a radioimmunologic technique for detection of the etiologic agents. Science 194:1428–1430

De Vries G (1982) Progress and problems in the eradication of EBL in the Netherlands. Curr Top Vet Med Anim Sci 15:510–513

Diglio CA, Ferrer JF (1976) Induction of syncytia by the bovine C-type leukemia virus. Cancer Res 36:1056–1067

Drescher B, Rössler H, Venker P, Wittmann W (1979) Purification and characterization of bovine leukemia virus DNA polymerase. Arch Geschwulstforsch 49:569–579

Driscoll DM, Olson C (1977) Bovine leukemia virus associated antigens in lymphocyte cultures. Am J Vet Res 38:1897–1898

Ferrer J (1982) Use of the radioimmunoassay in a program aimed at the evadication of bovine leukemia virus (BLV) infection from a high incidence herd. Curr Top Vet Med Anim Sci 15:532–542

Ferrer JF, Diglio CA (1976) Development of an in vitro infectivity assay for the C-type bovine leukemia virus. Cancer Res 36:1068–1073

Ferrer JF, Stock ND, Lin PS (1971) Detection of replicating C-type viruses in continuous cell cultures established from cows with leukemia: effect of the culture medium. J Natl Cancer Inst 47:613–621

Ferrer JF, Piper CE, Baliga V (1977) Diagnosis of BLV infection in cattle of various ages. In: Burny A (ed) Bovine leucosis: various methods of molecular virology. Comm Eur Communities, Luxembourg, pp 326–336

Fujimoto Y, Miller J, Olson C (1969) The fine structure of lymphosarcoma in cattle. Pathol Vet 6:15–29

Gerard-Marchand R, Hamlin I, Lennert K, Rilke F, Stanfeld AG, Van Unnik JAM (1974) Classification of non-Hodgkin's lymphomas. Lancet ii:406–408

Ghysdael J, Kettmann R, Burny A (1978) Translation of bovine leukemia virus genome information in heterologous protein synthesizing systems programmed with virion RNA and in cell-lines persistently infected by BLV. Ann Rech Vet 9:627–634

Ghysdael J, Kettmann R, Burny A (1979) Translation of BLV virion RNAs in heterologous protein synthesizing systems. J Virol 29:1087–1098

Gilden RV, Long CW, Hanson M, Toni R, Charman HP, Oroszlan S, Miller JM, Van der Maaten MJ (1975) Characteristics of the major internal protein and DNA-dependent DNA polymerase of bovine leukemia virus. J Gen Virol 29:305–314

Gràves DC, Ferrer JF (1976) In vitro transmission and propagation of the bovine leukemia virus in monolayer cell cultures. Cancer Res 36:4152–4159

Graves DC, Jones LV (1981) Early syncytium formation by bovine leukemia virus. J Virol 38:1055–1063

Graves DC, Diglio CA, Ferrer JF (1977) A reverse transcriptase assay for detection of bovine leukemia virus. Am J Vet Res 38:1739–1744

Guillemain B, Mamoun R, Levy D, Irgens K, Astier T, Parodi AL (1977) A quantitative study of early polykaryocytosis in mixed cultures of bovine leukemia virus infected bovine cells and mouse sarcoma virus infected cat cells. In: Burny A (ed) Bovine leucosis: various methods of molecular virology. Commission of the European Communities, Luxembourg, pp 347–356

Gupta P, Ferrer JF (1982) Expression of bovine leukemia virus genome is blocked by a non-immunoglobulin protein in plasma from infected cattle. Science 215:405–407

Hare WCD, Yang TJ, McFeely RA (1967) A survey of chromosome findings in 47 cases of bovine lymphosarcoma (leukemia), J Natl Cancer Inst 38:383–387

Hayward WS, Neel BG, Astrin SM (1981) Activation of a cellular *onc* gene by promoter insertion in ALV-induced lymphoid leukosis. Nature 290:475–480

Heuertz S, Hors-Cayla MC (1981) Cattle gene mapping by somatic cell hybridization study of 17 enzyme markers. Cytogenet Cell Genet 30:137–145

Hinuma Y, Nagata K, Hanaoka M, Nakai N, Matsumoto T, Kinoshita K, Shirawa S, Miyoshi I (1981) Adult T cell leukemia: antigen in an ATL cell-line and detection of antibodies to the antigen in human sera. Proc Natl Acad Sci USA 78:6476–6480

Horváth Z (1982) Experimental transmission of bovine leukosis to sheep (Clinicopathological considerations). Curr Top Vet Med Anim Sci 15:269–281

House JA, House C (1977) Studies on the bovine leukemia glycoprotein immunodiffusion test. In: Burny A (ed) Bovine leucosis: various methods of molecular virology. Commission of European Committees, Luxembourg, pp 237–246

Huber NL, Digiacomo RF, Evermann JF, Studer E (1981) Bovine leukemia virus infection in a large holstein herd: prospective comparison of production and reproductive performance in antibody-negative and antibody-positive cows. Am J Vet Res 42:1477–1481

Irgens K, Pinelli C, Guillemain B, Levy D, Parodi AL (1977) Early syntcytia formation induced by BLV producing cells in mixed cultures. In: Burny A (ed) Bovine leucosis: various methods of molecular virology. Communities, Luxembourg, pp 311–322

Jarrett WFH, Mackey LJ (1974) Bull WHO 50:1–142

Kaaden OR, Frenzel B, Dietzschold B, Weiland F, Mussgay M (1977) Isolation of a p15 polypeptide from bovine leukemia virus and detection of specific antibodies in leukemic cattle. Virology 77:501–509

Kaaden OR, Fischer W, Meerman A, Liebisch A (1982) Transmission of BLV by *Ixodes ricinus* ticks. Curr Top Vet Med Anim Sci 15:358–360

Kashmiri SVS, Mehdi R, Ferrer JF (1983) Detection, purification and characterization of two species of covalently closed circular proviral DNA molecules of bovine leukemia virus. J Virol 45:1172–1176

Kawakami TG, Moore AL, Theilen GH, Munn RJ (1970) Comparison of virus-like particles from leukotic cattle to feline leukosis virus. Bibl Haematol 36:471–475

Kenyon SJ, Piper CE (1977) Properties of density gradient-fractionated peripheral blood leukocytes from cattle infected with bovine leukemia virus. Infect Immun 16:898–903

Kenyon SJ, Gupta P, Ferrer JF (1982) Presence of the bovine leukemia virus (BLV) in milk of naturally infected cows. Curr Top Vet Med Anim Sci 15:289–297

Kettmann R, Portetelle D, Mammerickx M, Cleuter Y, Dekegel D, Galoux M, Ghysdael J, Burny A, Chantrenne H (1976) Bovine leukemia virus: an exogenous RNA oncogenic virus. Proc Natl Acad Sci USA 73:1014–1018

Kettmann R, Burny A, Cleuter Y, Ghysdael J, Mammerickx M (1978) Distribution of bovine leukemia virus proviral DNA sequences in tissues of animals with enzootic bovine leucosis. Leukemia Res 2:23–32

Kettmann R, Cleuter Y, Mammerickx M, Meunier-Rotival M, Bernardi G, Burny A, Chantrenne H (1980a) Genomic integration of bovine leukemia provirus: comparison of persistent lymphocytosis with lymph node tumor form of enzootic bovine leukosis. Proc Natl Acad Sci USA 77:2577–2581

Kettmann R, Marbaix G, Cleuter Y, Portetelle D, Mammerickx M, Burny A (1980b) Genomic integration of bovine leukemia provirus and lack of viral RNA expression in the target cells of cattle with different responses to BLV infection. Leukemia Res 4:509–519

Kettmann R, Deschamps J, Cleuter Y, Couez D, Burny A, Marbaix G (1982) Leukemogenesis by bovine leukemia virus: proviral DNA integration and lack of RNA expression of viral long terminal repeat and 3′ proximate cellular sequences. Proc Natl Acad Sci USA 79:2465–2469

Kettmann R, Westin EH, Marbaix G, Deschamps J, Wong-Staal F, Gallo RC, Burny A (1983a) Lack of expression of cellular homologues of retroviral *onc* genes in bovine tumors. In: Neth R et al (ed) Modern trends in human leukemia V. Springer, Berlin Heidelberg New York, pp 218–221

Kettmann R, Deschamps J, Couez D, Claustriaux JJ, Palm R, Burny A (1983b) J Virol 47:146–150

Kumar SP, Paul PS, Pomeroy KA, Johnson DW, Muscoplat CC, Van der Maaten MJ, Miller JM, Sorensen DK (1978) Frequency of lymphocytes bearing Fc receptors and surface immunoglobulins in normal persistent lymphocytotic and leukemic cows. Am J Vet Res 39:45–49

Langston A, Ferdinand GA, Ruppanner R, Theilen GH, Drlica S, Behymer D (1978) Comparison of production variables of bovine leukemia virus antibody-negative and antibody-positive cows in two California dairy herds. Am J Vet Res 39:1093–1098

Lennert K, Stein H, Kasserling E (1975) Cytological and functional criteria for the classification of malignant lymphomas. Br J Cancer 31 [suppl II]:29–43

Long CW, Henderson LE, Oroszlan S (1980) Isolation and characterization of low molecular-weight DNA binding proteins from retroviruses. Virology 104:491–496

Maas-Inderwiesen F, Albrecht A, Bause I, Osmers M, Schmidt FW (1978) Zum Einfluß der Leukosebekämpfung auf die Entwicklung der enzootischen Rinderleukose in Niedersachsen. Dtsch Tierärztl Wochenschr 85:309–314

Mammerickx M (1972a) Etude epizootologique sur les leucoses des animaux en Belgique de 1961 à 1970. Ann Med Vet 116:157–166

Mammerickx M (1972b) La transmission verticale et horizontale de la leucose bovine enzootique. Premiers résultats après cinq and d'expérimentation. Ann Med Vet 116:647–659

Mammerickx M (1973) L'examen nécropsique des bovidés leucémiques. Ann Med Vet 117:93–102

Mammerickx M, Cormann A, Burny A, Dekegel D, Portetelle D (1978) Eradication of enzootic bovine leukosis based on the detection of the disease by the gp51 immunodiffusion test. Ann Rech Vet 9:885–894

Mammerickx M, Portetelle D, Burny A (1981) Experimental cross-transmissions of bovine leukemia virus (BLV) between several animal species. Zentralbl Veterinaermed (B) 28:69–81

Mamoun RZ, Astier T, Guillemain B (1981) Establishment and propagation of a bovine leukemia virus-producing cell-line derived from the leukocytes of a leukemic cow. J Gen Virol 54:357–365

Mamoun RZ, Astier T, Guillemain B, Duplan JF (1983) Bovine lymphosarcoma: expression of BLV-related proteins in cultured cells. J Gen Virol 64:1895–1905

Marbaix G, Kettmann R, Cleuter Y, Burny A (1981) Viral RNA content of bovine leukemia virus-infected cells. Mol Biol Rep 7:135–138

Marín C, López N, Lozano O, Palencia L, España W, Castaños H, León H (1978) Epidemiology of bovine leukemia in Venezuela. Ann Rech Vet 9:743–746

Marín C, de Lopez N, de Alvarez M, Castaños H, España W, León A, Bello A (1982) Humoral spontaneous response to bovine leukaemia virus infection in zebu, sheep, buffalo and capybara. Curr Top Vet Med Anim Sci 15:310–320

McDonald HC, Ferrer JF (1976) Detection, quantitation and characterization of the major internal virion antigen of the bovine leukemia virus by radioimmunoassay. J Natl Cancer Inst 57:875–882

Miller JM, Olson C (1972) Precipitating antibody to an internal antigen of the C-type virus associated with bovine lymphosarcoma. J Natl Cancer Inst 49:1459–1462

Miller JM, Van der Maaten JM (1978a) Evaluation of an inactivated bovine leukemia virus preparation as an immunogen in cattle. Ann Rech Vet 9:871–877

Miller JM, Van der Maaten MJ (1978b) Infectivity tests of secretions and excretions from cattle infected by bovine leukemia virus. J Natl Cancer Inst 62:425–428

Miller JM, Miller LD, Olson C, Gillette KG (1969) Virus-like particles in phytohemagglutinin-stimulated lymphocyte culture with reference to bovine lymphosarcoma. J Natl Cancer Inst 43:1297–1305

Miller JM, Van der Maaten MJ, Schmerr MJ (1983) Vaccination of cattle with binary ethylenimine-treated bovine leukemia virus. Am J Vet Res 44:64–67

Muscoplat C, Johnson DW, Pomeroy KA, Olson JM, Larson VL, Stevens JB, Sorensen DK (1974) Lymphocyte surface immunoglobulin: frequency in normal and lymphocytotic cattle. Am J Vet Res 35:593–595

Ogura H, Paulsen J, Bauer H (1977) Cross neutralization of ovine and bovine C-type leukemia virus-induced syncytia formation. Cancer Res 37:1486–1489

Olson C, Kettmann R, Burny A, Kaja R (1981) Goat lymphosarcoma from bovine leukemia virus. J Natl Cancer Inst 67:671–675

Onuma M, Olson C, Baumgartener LE, Pearson LD (1975) An ether-sensitive antigen associated with bovine leukemia virus infection. J Natl Cancer Inst 55:1155–1158

Onuma M, Honma T, Mikami T, Ichijo S, Konishi T (1979) Studies on the sporadic and enzootic forms of bovine leukosis. J Comp Pathol 89:159–167

Onuma M, Hodatsu T, Yamamoto S, Higashihara M, Masu S, Izawa H, Mikami T (1983) Protection against bovine leukemia virus infection in sheep by vaccination. Am J Vet Res, in press

Oroszlan S, Copeland TD, Henderson LE, Stephenson JR, Gilden RV (1979) Amino-terminal sequence of bovine leukemia virus major internal protein: homology with mammalian type-C virus p30 structural proteins. Proc Natl Acad Sci USA 76:2996–3000

Oroszlan S, Sarngadharan MG, Copeland TD, Kalyanaraman VS, Gilden RV, Gallo RC (1982) Primary structure analysis of the major internal protein p24 of human type C T-cell leukemia virus. Proc Natl Acad Sci USA 79:1291–1294

Parodi AL, Mialot M, Crespeau F, Levy D, Salmon H, Nogues G, Gerard-Marchand R (1982) Attempt for a new cytological and cytoimmunological classification of bovine malignant lymphoma (BML) (lymphosarcoma). Curr Top Vet Med Anim Sci 15:561–572

Paul PS, Pomeroy KA, Castro AE, Johnson DW, Muscoplat CC, Sorensen DK (1977a) Detection of bovine leukemia virus in B-lymphocytes by the syncytia induction assay. J Natl Cancer Inst 59:1269–1271

Paul PS, Pomeroy KA, Johnson DW, Muscoplat CC, Handwerger BS, Soper FS, Sorensen DK (1977b) Evidence for the replication of bovine leukemia virus in the B-lymphocytes. Am J Vet Res 38:873–876

Payne GS, Bishop JM, Varmus HE (1981) Multiple arrangements of viral DNA and an activated host oncogene in bursal lymphomas. Nature 295:209–213

Poiesz BJ, Ruscetti FW, Gazdar AF, Bunn PA, Minna JD, Gallo RC (1980) Detection and isolation

of type-C retrovirus particles from fresh and cultured lymphocytes of a patient with cutaneous T-cell lymphoma. Proc Natl Acad Sci USA 77:7415–7419
Poiesz BJ, Ruscetti FW, Reitz MS, Kalyanaraman VS, Gallo RC (1981) Isolation of a new type-C retrovirus (HTLV) in primary uncultured cells of a patient with Sezary T-cell leukemia. Nature 294:268–271
Portetelle D, Mammerickx M, Beck F, Burny A, Cleuter Y, Dekegel D, Ghysdael J, Kettmann R, Chantrenne H (1977) Purification of BLV gp70 and BLV p24. Detection by radioimmunoassay of antibodies directed against these antigens. In: Burny A (ed) Bovine leucosis: various methods of molecular virology. Commission of the European Communities, Luxembourg, pp 131–152
Portetelle D, Bruck C, Burny A, Dekegel D, Mammerickx M, Urbain J (1978) Detection of complement-dependent lytic antibodies in sera from bovine leukemia virus-infected animals. Ann Rech Vet 9:667–674
Portetelle D, Bruck C, Mammerickx M, Burny A (1980) In animals infected by bovine leukemia virus (BLV) antibodies to envelope glycoprotein gp51 are directed against the carbohydrate moiety. Virology 105:223–233
Portetelle D, Bruck C, Mammerickx M, Burny A (1983) Use of monoclonal antibody in an ELISA test for the detection of antibodies to bovine leukaemia virus. J Virol Meth 6:19–29
Ressang AA, Mastenbroek N, Quak J, Van Griensven LJ, Calafat J, Hilgers J, Hageman P, Souissi T, Swen S (1974) Studies on bovine leukemia. I. Establishment of type-C virus producing cell lines. Zentralbl Veterinaermed (B) 21:602–617
Schmerr MJ, Miller JM, Van der Maaten MJ (1981) Antigenic reactivity of a soluble glycoprotein associated with bovine leukemia virus. Virology 109:431–434
Schmidt FW (1982) Results and observations of an EBL eradication program based on AGID diagnosis and culling of reactors. Curr Top Vet Med Anim Sci 15:491–497
Seiki M, Hattori S, Yoshida M (1982) Human adult T-cell leukemia virus: molecular cloning of the provirus DNA and the unique terminal structures. Proc Natl Acad Sci USA 79:6899–6902
Sentsui H, Thorn RM, Kono Y, Ferrer JF (1982) Haemagglutination by bovine leukaemia virus. J Gen Virol 59:83–89
Stock ND, Ferrer JF (1972) Replicating C-type virus in phytohemagglutinin-treated buffy-coat cultures of bovine origin. J Natl Cancer Inst 48:985–996
Straub O (1982) Transmission studies from leukemic cattle to sheep using secretions, excretions, breath and skin scrapings. Curr Top Vet Med Anim Sci 15:299–308
Straub OC (1978) Preliminary results of a new sanitation program for the eradication of enzootic bovine leukosis. Ann Rech Vet 9:895–898
Takashima I, Olson C, Driscoll DM, Baumgartener LE (1977) B- and T-lymphocytes in three types of bovine lymphosarcoma. J Natl Cancer Inst 59:1205–1209
Theilen GH, Miller JM, Higgins J, Ruppaner RN, Garrett W (1982) Vaccination against bovine leukemia virus infection. Curr Top Vet Med Anim Sci 15:547–559
Uckert W, Wunderlich V (1979) Proteins of bovine leukemia virus: p15 is the major phosphoprotein. Acta Biol Med Ger 38:K42
Uckert W, Wunderlich V, Ghysdael J, Portetelle D, Burny A (1984) Bovine leukemia virus (BLV – structural model based on chemical cross-linking studies. Virology, in press
Van der Maaten MJ, Miller JM (1976a) Serological evidence of transmission of bovine leukemia virus to chimpanzees. Vet Microbiol 1:351–357
Van der Maaten MJ, Miller JM (1976b) Replication of bovine leukemia virus in monolayer cell cultures. Bibl Haematol 43:360–362
Van der Maaten MJ, Miller JM (1977) In: Cold Spring Harbor Lab (ed) Origins of human cancer. Cold Spring Harbor Laboratory, New York, pp 1223–1234
Van der Maaten MJ, Miller JM, Boothe AD (1974) Replicating type-C virus particles in monolayer cell cultures of tissue from cattle with lymphosarcoma. J Natl Cancer Inst 52:491–494
Weiland F, Straub OC (1975) Frequency of surface immunoglubulin-bearing blood lymphocytes in cattle affected with bovine leukosis. Res Vet Sci 19:100–102
Wilesmith JW, Straub OC, Lorenz RJ (1978) Untersuchungen zur iatrogenen Übertragung des Virus der Rinderleukose. Tierärztl Umschau 33:519–523
Wittmann W, Urbaneck D (1969) Leukosen des Rindes. In: Röhren H (ed) Handbuch der Virusinfektionen bei Tieren, vol V/1. Fischer, Jena
Yoshikawa T, Yoshikawa H, Takeuchi S (1972) Bovine leukosis cluster in the Towada district of Japan. Kitasato Arch Exp Med 45:193–204

Spleen Focus-Forming Virus: Relationship of an Altered Envelope Gene to the Development of a Rapid Erythroleukemia

S. RUSCETTI and L. WOLFF

1	Introduction	21
2	Effects of SFFV on Hematopoietic Cells	22
2.1	Friend $SFFV_P$	22
2.2	Friend $SFFV_A$	24
2.3	Rauscher SFFV and Cas SFFV	24
2.4	Multiple Stages of SFFV-Induced Disease	25
3	General Organization of the SFFV Genome	25
4	Genetic Elements of SFFV Required for Pathogenicity	27
4.1	Early Studies	27
4.2	Studies with Subgenomic Fragments and Genetically Engineered Mutants	28
4.2.1	Studies with $SFFV_P$	28
4.2.2	Studies with $SFFV_A$	29
4.3	Studies of Spontaneous Mutants of SFFV	29
5	Comparison of the Envelope Genes of SFFV and MCF Viruses	29
6	Unique Structural Characteristics of the SFFV Envelope Gene Product	32
6.1	Fusion Glycoprotein	32
6.2	Carboxyl Terminus	33
6.3	Glycosylation	34
7	Molecular Basis for the $SFFV_P$ and $SFFV_A$ Phenotypes	35
8	Conclusions and Speculations	36
References		39

1 Introduction

In the 26 years since "Friend virus" was discovered (FRIEND 1957), we have learned a considerable amount about this virus complex and the role of the various components in the development of erythroleukemia in the mouse. The biological and molecular cloning of these viral components has been pivotal in carrying out these studies. Separation in rodent fibroblasts of Friend murine leukemia virus (F-MuLV) and the defective spleen focus-forming virus (SFFV) has led to the conclusion that SFFV is responsible for the rapid erythroleukemia occurring in adult mice, while the role of F-MuLV in the adult is to act as a helper for the replicative functions that the defective SFFV lacks. Molecular cloning of SFFV as well as the isolation of biological mutants of SFFV has allowed for the localization of the region of the viral genome involved in patho-

Laboratory of Genetics, National Cancer Institute, Bethesda, MD 20205, USA

Current Topics in Microbiology and Immunology, Vol. 112

genicity. This region includes almost exclusively the envelope gene of SFFV. The focus of this review, therefore, will be on studies of the envelope gene of various SFFVs and their translational products.

2 Effects of SFFV on Hematopoietic Cells

Any data on the genes and translational products of SFFV must relate to the effects of SFFV on hematopoietic cells. A variety of studies to determine these have been carried out both in vivo and in vitro. Since SFFV is a defective virus, all studies must be carried out in the presence of a helper virus. A number of studies have shown, however, that it is the SFFV component in the viral complex that is responsible for the hematopoietic changes induced in adult mice (TROXLER et al. 1980; HANKINS and KRANTZ 1980; MACDONALD et al. 1980a, 1981; FAGG and OSTERTAG 1982). Although the effects of all preparations of Friend SFFV, unlike the majority of murine leukemia viruses, are specific for the erythroid lineage, variants of SFFV derived from the same original virus stock can be distinguished by their ability to induce polycythemia (designated $SFFV_P$) or a mild anemia (designated $SFFV_A$).

2.1 Friend $SFFV_P$

Infection of mice with the $SFFV_P$ viral complex leads to a rapid (within 30 h) and massive proliferation in the spleen of hyperbasophilic proerythroblast-like cells (Friend cells) (TAMBOURIN and WENDLING 1971). If a low dose of virus is given, this erythroid hyperplasia can be seen as distinct splenic foci in 9–10 days (AXELRAD and STEEVES 1964; MIRAND et al. 1968a). By 3 weeks postinfection, the spleen and liver become greatly enlarged and, due to active erythropoiesis, polycythemia develops (MIRAND et al. 1961; TAMBOURIN and WENDLING 1971). The disease proves fatal in 3–6 weeks.

The effects of $SFFV_P$ on erythropoiesis are dramatic. Erythroid differentiation, which is normally under control of the hormone erythropoietin, Epo, is seemingly uncontrolled in $SFFV_P$-infected mice. These mice continue to produce mature red blood cells under conditions where Epo levels are undetectable and erythropoiesis should cease (MIRAND 1967; MIRAND et al. 1968b; SASSA et al. 1968; TAMBOURIN and WENDLING 1971). When hematopoietic cells from these mice are tested in in vitro colony assays which detect Epo-responsive erythroid precursor cells at a relatively late stage of differentiation (designated CFU-E), a large increase in these colonies is seen, beginning approximately 3 days postinfection (HOROSZEWICZ et al. 1975; LIAO and AXELRAD 1975; HANKINS and KRANTZ 1975). While normal cells require Epo for these colonies to form, cells from $SFFV_P$-infected mice can form CFU-E in the absence of exogenously added Epo.

Further information concerning the effects of $SFFV_P$ on erythropoiesis has come from infecting hematopoietic cells in vitro. When normal bone marrow cells are infected in vitro with $SFFV_P$ that has been pseudotyped with any of a number of helper MuLVs and then grown in semisolid media without Epo, large clusters of erythroid "bursts" form after 5 days (HANKINS et al. 1978). These "bursts" are probably the precursors of the cells responsible for the increase in the number of CFU-E seen in the animal, which take 3 days in the mouse and 2 days in culture to develop. In order to distinguish these day-5 virus-induced erythroid bursts from the normal erythroid bursts (designated BFU-E) that appear after 7–10 days in culture, they have been designated vBFU-E (KOST et al. 1979). While normal cells require the addition of Epo to the culture in order to form BFU-E, SFFV-infected cells can form vBFU-E without the addition of exogenous Epo. Since colony assays are routinely carried out in the presence of fetal calf serum, which may contain some endogenous Epo, it is not known whether erythroid differentiation of $SFFV_P$-infected cells is completely independent of Epo or extremely sensitive to low levels of the hormone. Unfortunately, colony assays carried out under serum-free conditions have given conflicting results (ROSSI and PESCHLE 1980; KOST et al. 1980; MAC DONALD et al. 1981).

SFFV has also been used to infect long-term bone marrow suspension cultures. While early studies, carried out under conditions that select a disproportionate number of granulocytes and stem cells, indicated that such infection could give rise to pluripotent stem cell lines (DEXTER et al. 1979) or granulocytic cell lines (DEXTER et al. 1977; GREENBERGER et al. 1979), more recent work (DEXTER et al. 1981), using conditions that favor erythropoiesis, has shown that $SFFV_P$ can induce changes in long-term bone marrow suspension cultures similar to those seen after $SFFV_P$ infection in vivo, such as the accumulation of a large number of cells that are capable of forming CFU-E in the absence of exogenous Epo.

Although SFFV may infect a variety of hematopoietic cells, the cell upon which the virus has its effect, i.e., the target, appears to be an Epo-responsive erythroid cell (TAMBOURIN and WENDLING 1971, 1975). Treatment of bone marrow cells in vivo or in vitro with actinomycin D, which specifically decreases the number of Epo-responsive erythroid precursor cells (CFU-E), significantly reduces the number of target cells available for the virus (TAMBOURIN and WENDLING 1971; OPITZ and SEIDEL 1978; KOST et al. 1979). Also, any treatment that will lead to an increase in the Epo-responsive cell compartment will increase the number of target cells for SFFV (STEEVES et al. 1969; TAMBOURIN and WENDLING 1971; FREDRICKSON et al. 1975; KOST et al. 1979, 1981).

Studies by KOST et al. (1979, 1981), using cell separation techniques, have suggested that the target cell for $SFFV_P$ is a relatively late committed erythroid precursor cell that is located in a compartment between the day-8 BFU-E and the day-2 CFU-E. Their studies also showed that the target cell for SFFV must be in a state of active cellular DNA synthesis in order for the virus to have its effect. This may be necessary for the proper integration of the SFFV genome into the host cellular DNA. This lends support to the idea that the target cell is closer to a CFU-E, which is a rapidly cycling cell, than to a primitive BFU-E or a spleen colony-forming unit (CFU-S), the majority of which are not in cell cycle (GREGORY et al. 1973). Since erythroid differentiation is likely to be a delicately balanced process involving stimulators and inhibitors of differentiation, it is possible that SFFV may indirectly have effects on erythropoiesis by altering the cells producing these factors. Several studies support this idea. PESCHLE et al. (1981) have reported that SFFV infection leads to both an increase in the production of and hypersensitivity of cells to burst-promoting activity (BPA), a factor which is thought to increase the proliferation and differentiation of early erythroid precursor cells (ISCOVE 1978; WAGEMAKER 1978). Conversely, JOHNSON et al. (1982) showed that macrophage-like cells from SFFV-infected mice produce high levels of a factor which inhibits normal erythroid differentiation but stimulates differentiation of erythroid precursor cells from SFFV-infected mice. In either case, the result may be an increase in the number of suitable target cells for the virus.

2.2 Friend $SFFV_A$

Like mice infected with $SFFV_P$, $SFFV_A$-infected mice develop greatly enlarged spleens and livers due to a massive infiltration of hyperbasophilic proerythroblasts (FRIEND 1957; METCALF et al. 1959). Foci can be detected on the surface of the spleen in 9–10 days postinfection (TROXLER et al. 1980; MACDONALD et al. 1980a), but these foci are not as distinct as those induced by $SFFV_P$. $SFFV_A$-infected mice, in contrast to $SFFV_P$-infected mice, do not become polycythemic but develop a mild anemia (STEINHEIDER et al. 1979). Unlike $SFFV_P$-infected mice, erythropoiesis in mice infected with $SFFV_A$ is sensitive to normal physiological controls (TAMBOURIN et al. 1979). Hematopoietic cells from these mice show an increase in the number of cells capable of forming CFU-E in culture (STEINHEIDER et al. 1979; MACDONALD et al. 1980a), but these colonies, as compared to those from $SFFV_P$-infected mice, are not able to form in the absence of exogenous Epo. Like infection with $SFFV_P$, infection of normal bone marrow cells in vitro with $SFFV_A$ also leads to the development of day-5 erythroid "bursts" (vBFU-E) (HANKINS and TROXLER 1980; KOURY et al. 1982). However, the $SFFV_A$ bursts do not differentiate into mature red blood cells unless a small amount of Epo is added, an amount too low to induce differentiation of uninfected erythroid precursor cells.

Although $SFFV_A$ appears to interfere less than $SFFV_P$ with erythroid differentiation, both viruses increase the proliferative capacity of erythroid precursor cells and enable them to differentiate under conditions that are suboptimal for uninfected cells. Thus, the target cell for $SFFV_A$ is probably very close to or identical to that for $SFFV_P$. The different phenotypes seen may be due to qualitative differences in viral gene products or to differences in subcellular location of these products in the target cells (see below).

2.3 Rauscher SFFV and Cas SFFV

In 1962, Rauscher isolated a virus which induced a rapid erythroproliferative disease very similar to the disease induced by Friend $SFFV_A$. This virus complex was shown to contain a defective spleen focus-forming virus component (PLUZNIK and SACHS 1964) and has the same biological effects on hematopoietic cells as Friend $SFFV_A$, resulting in an increase in the number of day-5 erythroid bursts that still require a small amount of Epo for differentiation (HANKINS and TROXLER 1980). Interestingly, unlike Friend SFFV, a polycythemia-inducing variant of Rauscher SFFV has never been isolated.

Recently, LANGDON et al. (1983a, b) isolated another murine erythroleukemia-inducing virus that is very similar to Friend $SFFV_A$ and Rauscher SFFV. This virus, which was shown to contain a defective SFFV component designated Cas SFFV, was isolated from mice receiving a spleen homogenate from lymphomatous mice previously infected with the wild mouse ecotropic virus Cas-Br-M. In vitro infection of bone marrow cells with Cas SFFV leads to the development of day-5 erythroid bursts in the presence of low concentrations of Epo (LANGDON et al. 1983b).

2.4 Multiple Stages of SFFV-Induced Disease

While it is possible to establish transplantable cell lines from diseased tissues of SFFV-infected mice late in the course of the disease (FRIEND and HADDAD 1960), the majority of the proliferating cells early in the disease have a limited self-renewal capacity and fail to grow in syngeneic donor mice or as established cell lines in vitro. It was, therefore, proposed (TAMBOURIN et al. 1979, 1980; LEVY et al. 1979) that the disease induced by SFFV consists of multiple stages: (a) an early stage in which the accumulating erythroid precursor cells have a finite life span but the size of the compartment of erythroid precursor cells is uncontrolled and (b) a late stage (>3 weeks postinfection) which represents the proliferation of truly malignant cells that are capable of autonomous and extensive proliferation and tumorigenic growth in vivo. The latter cells can be identified by their ability to form spleen colonies in vivo in genetically anemic Sl/Sl^d mice (MAGER et al. 1980), to form macroscopic colonies in vitro (MAGER et al. 1981), and to grow in the omentum of mice after intraperitoneal injection (WENDLING et al. 1981). It is not known whether the cells proliferating in the first stage are progenitors of the malignant cells in the latter stage or whether the two populations of cells arise independently. Most cell lines derived from the malignant stage are more or less blocked in their ability to differentiate, although they can be induced to differentiate with certain chemicals such as dimethyl sulfoxide (FRIEND et al. 1971). Both SFFV and helper viral proteins can be detected at all stages of the disease, although high levels of a transformation-related protein, p53 (for review see KLEIN 1982), can be detected only in the malignant stage of the disease (RUSCETTI and SCOLNICK 1983).

While it is clear that certain genes contained in SFFV are necessary for the initiation of the early disease (see below), the role of these genes or any others in the malignant stage of the disease is presently unknown.

3 General Organization of the SFFV Genome

Once the spleen focus-forming virus was biologically cloned free of its helper virus (TROXLER et al. 1977a; BERNSTEIN et al. 1977), it was possible for the first time to study its genome and compare it with that of its natural ecotropic helper, Friend MuLV. It should be mentioned that studies on Friend $SFFV_P$ have been carried out on several different strains of the virus, all derived from the original Friend virus stock. Although differences can be found in the genomes and gene products of these viruses, they have identical effects on erythropoiesis.

The spleen focus-forming virus appears to have arisen by recombination between its natural helper virus and endogenous mouse genetic sequences (TROXLER et al. 1977b, c). The latter sequences are highly related to the envelope (*env*) genes present in mink-cell focus-inducing (MCF) viruses, which have also arisen by recombination of ecotropic murine leukemia viruses with *env* genes endogenously present in the mouse DNA (for review see FAMULARI 1983). Thus, SFFV and MCF viruses appear to have a common origin. Due to their acquired *env* genes, MCF viruses bind to unique receptors on mouse cells (REIN 1982) and have been implicated in a variety of MuLV-induced leukemias (for review see FAMULARI 1983). While most MCF viruses are associated with lymphoid leukemias, Friend MCF virus has been implicated (RUSCETTI et al. 1981a) in an F-MuLV-induced erythroproliferative disease that is quite similar, yet distinct, from that induced by SFFV (TROXLER and SCOLNICK 1978).

Consistent with its replication defectiveness, $SFFV_P$ was shown to contain a 32S RNA (EVANS et al. 1979), in contrast to the 38S RNA for F-MuLV, and to lack sequences capable of encoding a functional RNA-dependent DNA polymerase (BERNSTEIN et al. 1977). SFFV RNA contains sequences related to the 5′ end of the *gag* gene of F-MuLV (BERNSTEIN et al. 1977; BARBACID et al. 1978; RUSCETTI et al. 1980), but the amount appears to vary from strain to strain, resulting in the production of a 15000- to 45000-dalton protein as compared with the 65000-dalton prototype gag precursor. It also contains *env* gene-related sequences that encode a 52000- to 55000-dalton envelope glycoprotein (RACEVSKIS and KOCH 1978; IKAWA et al. 1978; RUSCETTI et al. 1978, 1979; DRESLER et al. 1979) that is immunologically related to the gp70s of MCF viruses (RUSCETTI et al. 1978, 1979). A 21S subgenomic RNA has been shown to code for this SFFV envelope protein (RUSCETTI et al. 1980; BILELLO et al. 1980; YOSHIDA and YOSHIKURA 1980). The *env* gene of $SFFV_P$ also contains information that results in the inefficient processing of the primary *env* gene product to a higher molecular weight form which appears on the cell surface (RUSCETTI et al. 1979, 1980; RUTA and KABAT 1980). Thus, Friend $SFFV_P$ appears to have arisen by deletion and recombination between F-MuLV and envelope (or other) genetic sequences endogenously present in the mouse.

The genome of $SFFV_A$, which is 30S in size, is slightly smaller than $SFFV_P$, but like $SFFV_P$ it also contains sequences that are highly related to the envelope gene of MCF viruses (EVANS et al. 1980; TROXLER et al. 1980). $SFFV_A$ encodes a 52000- to 55000-dalton envelope glycoprotein and a 45000-dalton gag-related protein (TROXLER et al. 1980; MACDONALD et al. 1980a).

Like other murine retroviruses, the integrated SFFV genome contains two long terminal repeats (LTRs), 514 bp each, at either end (CLARK and MAK 1982). These LTRs contain recognizable transcriptional regulatory sequences and short terminal inverted repeats, and include a functional promoter region that can activate the expression of downstream coding sequences.

Unlike other defective, acute leukemia viruses, a unique cellular sequence has not been detected in the genome of SFFV. Although SFFV does contain sequences that can be detected in normal mouse DNA (Chattopadhyay et al. 1982), these are viral envelope gene sequences related to those that have been acquired by MCF viruses. Thus, the unique properties of SFFV appear to reside in a combination of genetically altered murine leukemia virus sequences (see below).

Once SFFV had been molecularly cloned and shown to be biologically active when assayed in the presence of helper virus (Linemeyer et al. 1980), it enabled studies to determine which areas of the SFFV genome are crucial for the pathogenicity of the virus.

4 Genetic Elements of SFFV Required for Pathogenicity

4.1 Early Studies

As mentioned above, SFFV is known to encode two proteins: a 52–55K viral envelope-related protein and a 15–45K gag-related protein. Early studies suggested that the envelope protein was the most important of the two. For example, while strains of SFFV vary in terms of the gag proteins they express, with some expressing only p15 and others expressing a 45K gag protein containing p15, p12, and some p30 determinants, they all produce a highly conserved envelope protein that contains determinants related to MCF gp70 (Ruscetti et al. 1980). Also, while some erythroleukemia cell lines derived from mice infected with SFFV fail to express SFFV gag proteins, they all continue to express the SFFV envelope glycoprotein (Anand et al. 1981). The failure of some strains of SFFV to encode gag proteins may indicate that these proteins are not involved in the essential functions of SFFV. In contrast, the fact that gp52 is encoded by all strains of SFFV and that high levels of this protein are consistently found in all lines suggested that gp52 is involved in pathogenicity.

Other data supporting the importance of the *env* gene in oncogenicity came from comparisons of SFFV and its ecotropic parent by oligonucleotide fingerprinting. It was found that the only oligonucleotides unique to Friend $SFFV_P$ were located at the 3′ end of the genome, the area encoding the envelope proteins (Evans et al. 1979, 1980). Friend $SFFV_A$ RNA was also shown to contain a unique sequence in the envelope region (Evans et al. 1980). When Rauscher SFFV was compared with its ecotropic parent, the only difference was in the envelope gene, which encodes a 54000-dalton glycoprotein which is highly related to that encoded by Friend SFFVs (Ruta and Kabat 1980; Bestwick et al. 1983). In fact, Rauscher SFFV, unlike Friend SFFVs, contains functional *gag* and *pol* genes and is indistinguishable from Rauscher MuLV except for its *env* gene.

4.2 Studies with Subgenomic Fragments and Genetically Engineered Mutants

4.2.1 Studies with $SFFV_P$

Linemeyer et al. (1981) molecularly cloned a subgenomic DNA fragment of $SFFV_P$ (2.4 kbp in length) which included 2.0 kbp from the 3′ end of SFFV (which encode the viral envelope protein), the long terminal repeat, and 0.4 kbp from the 5′ end of SFFV (which do not encode any known viral proteins). The fragment was shown to contain all of the MCF-related *env* gene oligonucleotides previously mapped by Evans et al. (1979). When analyzed by a two-stage cotransfection assay, utilizing helper virus DNA derived from Friend, Moloney, or amphotropic MuLV, this subgenomic SFFV fragment was shown to retain full biological activity. They were subsequently able to clone a second 2.4-kbp fragment from the first by eliminating 0.6 kbp from the 3′ end (Linemeyer et al. 1982), and this smaller fragment still retained full biological activity. Mice infected with either subgenomic fragment were shown to express the SFFV envelope glycoprotein, gp52. In addition, both fragments were active in vitro in promoting the development of Epo-independent vBFU-E (D. Hankins, personal communication). These data indicated that the amount of SFFV information required for the induction of disease is less than 50% of the viral genome and ruled out the possibility that the *gag* gene is crucial for pathogenicity.

The pathogenic region was further defined when Linemeyer et al. (1982) produced mutations at random sites in the full-length SFFV genome and then cloned and analyzed nine mutant genomes for biological activity. DNA from clones which had a deletion in the *gag* gene retained the full biological activity of SFFV, as did clones with deletions covering most or all of the LTR region. Clones of SFFV DNA with deletions extending leftward into the *env* gene sequences, including one clone with deletions only in the *env* gene, were negative for biological activity. These results are consistent with the hypothesis that the *env* sequences of SFFV are necessary for the initiation of disease. Since the biologically active virus rescued from the cotransfections is probably formed after recombination of the SFFV sequences with those of the helper, it is possible that helper viral sequences act in conjunction with SFFV *env* sequences to cause disease. Any contribution from the helper sequences, however, would have to be nonspecific, since a variety of helpers, including those not associated with erythroid disease, can be used to rescue SFFV. Transfection of molecularly cloned fragments of SFFV DNA into hematopoietic cells in the absence of helper virus will be necessary to rule out a role for the helper virus.

By combining the results obtained with subgenomic fragments and with mutant DNAs, one can now define a 1.5-kbp segment of SFFV DNA which contains the only unique SFFV information essential for the induction of disease. Approximately 1.4 kbp of DNA would be required to encode a protein of 46K, which is the size of the unglycosylated SFFV gp52 (Schultz et al. 1980; Ikawa et al. 1980). Thus, the biologically active 1.5-kbp segment of DNA contains only slightly more information than that necessary to encode the recombinant gp52.

4.2.2 Studies with $SFFV_A$

The proviral DNA of $SFFV_A$ has also been molecularly cloned and from this DNA a 3.2-kbp subgenomic fragment, representing 2 kbp of the 3′ end of the viral genome (and containing the envelope gene) and two LTRs, was prepared (KAMINCHIK et al. 1982). When analyzed by the two-stage cotransfection assay, utilizing helper virus DNA derived from Moloney MuLV, this subgenomic fragment of $SFFV_A$ was shown to retain full biological activity, including the phenotype that distinguishes it from $SFFV_P$: the inability to cause polycythemia and the failure to lead to the development of fully hemoglobinized day-5 vBFU-E in the absence of Epo. Thus, the biological differences between $SFFV_P$ and $SFFV_A$ reside in their envelope genes (see below).

4.3 Studies of Spontaneous Mutants of SFFV

RUTA et al. (1983) have isolated transmissible F-$SFFV_P$ mutants with abnormalities in their envelope genes. These mutants, which have not lost any genetic information necessary for virus replication, represent deletions or changes in three nonoverlapping regions of the envelope gene and all are nonleukemogenic. They encode envelope glycoproteins that are smaller or larger than that encoded by wild-type virus and, unlike the wild-type protein, are defective in their processing to the plasma membranes. The fact that different nonoverlapping mutations in the envelope gene can all eliminate pathogenicity provides strong evidence that the entire, properly folded gene product may be required for pathogenicity.

Several transmissible mutants have also been isolated for Rauscher SFFV (MACHIDA et al. 1984). These mutants contain altered envelope genes and affect nonoverlapping domains of the envelope glycoprotein. They have dramatically reduced leukemogenicities compared with wild-type virus and, like Friend SFFV mutants, they appear to encode envelope proteins that are defective in their processing to the plasma membrane form. Thus, the plasma membrane form of the SFFV envelope glycoproteins may be important for initiation of erythroid disease.

During the course of cloning spontaneous mutants of SFFV, RUTA et al. (1983) also isolated a large number of SFFV clones that encode wild-type envelope glycoproteins. Unlike clones with mutant *env* genes, all of these mutants are pathogenic. Thus *env* gene mutations and loss of pathogenicity cosegregate during virus cloning, suggesting that they are causally related.

5 Comparison of the Envelope Genes of SFFV and MCF Viruses

Since the envelope genes of both Friend and Rauscher SFFVs appear to be crucial for the pathogenicity of these viruses, it is important to understand the structure of their *env* genes and gene products in hopes that such information may give us clues as to why the expression of these genes results in alterations of erythropoiesis.

As mentioned earlier, one of the first observations that was made regarding the genetic composition of SFFV was its relationship to mink-cell focus-inducing (MCF) viruses. Studies carried out by TROXLER et al. (1977b, c), using liquid hybridization techniques, indicated that SFFV was a recombinant virus containing parental F-MuLV sequences as well as sequences highly related to the *env* gene sequences present in MCF viruses. This was found to be true not only for Friend $SFFV_P$ but also for Friend $SFFA_A$ (TROXLER et al. 1980; EVANS et al. 1980), Rauscher SFFV (BESTWICK et al. 1983), and Cas SFFV (LANGDON et al. 1983a, b). Thus, all SFFVs appear to have envelope genes which are highly related to those contained in MCF viruses.

The location of the MCF virus-related sequences within the envelope region of SFFV was determined by comparisons of SFFV with MCF viruses at both the genetic level and the protein level. It had previously been shown that MCF viruses contain unique envelope genes and encode envelope glycoproteins that are distinct from those encoded by ecotropic MuLVs (for review see FAMULARI 1983). When SFFV RNA was compared by oligonucleotide fingerprinting with cDNAs from F-MuLV and Friend MCF virus, it was found that out of a total of 24 RNase T_1-resistant oligonucleotides in SFFV RNA, 3 could be recovered from hybrids formed with Friend MCF virus but not from hybrids with F-MuLV (EVANS et al. 1980). These sequences mapped to the 3′ end of the SFFV genome and were highly related to sequences found in Friend MCF virus. In addition, analyses performed by BOSSELMAN et al. (1980) and M. Gonda (personal communication) of heteroduplexes between F-MuLV and SFFV showed that the region of substitution was 0.8–0.9 kb in size and was located from 1.73 to 2.53 kbp from the 3′ end of the genome.

Additional evidence that the substituted genetic information in the SFFV genome is located in the envelope region and is highly related to the MCF envelope gene has come from analysis of Friend SFFV translational products. It was initially found that cells nonproductively infected with SFFV expressed a protein that could compete in a radioimmunoassay that detects the gp70s of a variety of MCF viruses but not ecotropic viruses (RUSCETTI et al. 1978). Later, antisera containing antibodies capable of precipitating MCF, but not ecotropic, gp70s were shown to precipitate a glycoprotein of 52000- to 55000 daltons from fibroblasts non-productively infected with Friend $SFFV_P$ (RUSCETTI et al. 1979), Friend $SFFV_A$ (TROXLER et al. 1980), and Rauscher SFFV (RUTA and KABAT 1980). Recently, Cas SFFV, derived from wild mouse ecotropic virus, was also shown to encode an MCF gp70-related protein of 50000 daltons (LANGDON et al. 1983b). The relationship of the SFFV glycoprotein to MCF gp70 was further supported at the biochemical level by the finding of common peptides in tryptic peptide fingerprints of [^{35}S]methionine-labeled SFFV gp52 and MCF gp70 (RUSCETTI et al. 1979).

Localization of the MCF gp70-related region within the SFFV gp52 molecule itself was accomplished through the use of monoclonal antibody to gp52 in conjunction with partial protease digestion. Both Friend MCF gp70 and gp52 contain a 23K V8 protease fragment that is precipitated by a monoclonal anti-

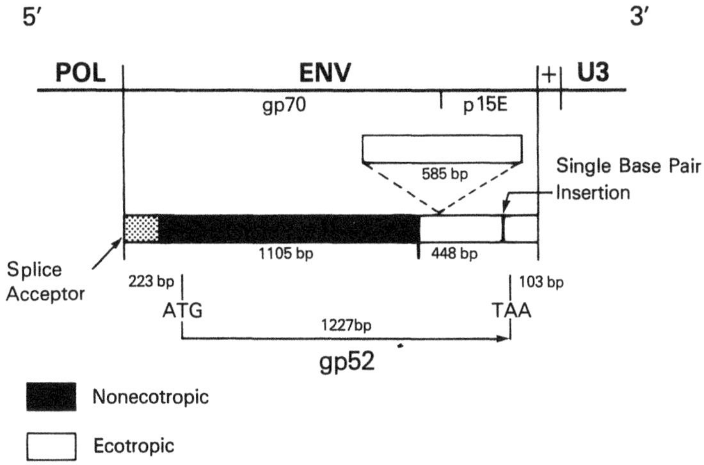

Fig. 1. Genetic structure of the *env* region of SFFV as determined by nucleotide sequence analysis. The *open areas* are regions that are homologous to the envelope of F-MuLV, the putative ecotropic parent. The *black area* is nonhomologous to F-MuLV, but highly related to xenotropic and MCF viruses. The *shaded area* is a region of the envelope where the sequence relationship to F-MuLV was not determined. The open reading frame that codes for gp52 begins at the ATG codon and ends at the termination codon TAA. The potential splice site for the envelope is indicated as well as the location of a single base pair insertion at the 3′ end that produces a shift in the reading frame

body to SFFV gp52 (WOLFF et al. 1984). The 23K fragment in Friend MCF gp70 was localized to the N-terminus (WOLFF et al. 1982), suggesting that the MCF-related region of gp52 was also at the N-terminus. The 23K peptides from Friend MCF gp70 and pg52 have produced tryptic peptide fingerprints which are indistinguishable (WOLFF et al. 1984).

Recently published nucleotide sequences of the envelope genes of several SFFVs (AMANUMA et al. 1983; WOLFF et al. 1983; CLARK and MAK 1983) have made it possible to make detailed comparisons of these with a number of MCF viral envelope genes. For three isolates of the polycythemia-inducing variants of SFFV it has been found that the MCF-related region of the SFFV *env* gene encodes more than two-thirds of the gp52 molecule. This includes the leader peptide and amino terminus of the glycoprotein itself (see Fig. 1). For all of the isolates, the amino acid homology with Moloney MCF (BOSSELMAN et al. 1982), AKR MCF 247 (HOLLAND et al. 1983), or Friend MCF (KOCH et al. 1984) gp70 in the analogous amino-terminal region is greater than 90–95%. Interestingly, there is a 67 amino acid region (starting at amino acid residue number 60 from the amino terminus) which is identical for all of these SFFVs and MCF viral *env* genes. The fact that this amino-terminal domain is so highly conserved in the envelope proteins from this group of pathogenic viruses may indicate that it serves a role in maintenance of the protein structure or as an important sequence in a functionally active site of the molecule. The nucleotide sequence of the 5′ end of the *env* gene of Friend $SFFV_A$ (L. Wolff, unpublished data) also shows extensive homology with the corresponding region of MCF viruses.

6 Unique Structural Characteristics of the SFFV Envelope Gene Product

There are several features of the SFFV *env* gene product, gp52, which are unique only to this retroviral protein. These features are obvious when gp52 is compared with *env* translation products of other retroviruses (for review see WEISS et al. 1982). The primary translation product of the prototype envelope gene is a polyprotein containing gp70 and Pr15E. This polyprotein, which can range in size from 80000 to 90000 daltons, is glycosylated with high mannose-containing oligosaccharides. Cleavage of this precursor into gp70 and Pr15E is temporally linked to modification of mannose-rich sugar residues on gp70 and the addition at these sites of complex terminal sugars. Further processing of Pr15E involves cleavage of Pr15E into p15E and an R peptide. As shown below, SFFV gp52 differs in a number of ways from this prototype envelope protein due to specific deletions and insertions in its *env* gene.

6.1 Fusion Glycoprotein

One unusual property of the SFFV *env* gene product is the covalent linkage of partial gp70 and p15E domains. Due to a large (585 base pair) deletion in the middle of the SFFV envelope gene, which includes the cleavage site between gp70 and Pr15E, the amino-terminal gp70 and carboxy-terminal p15E sequences are joined and remain covalently linked (see Fig. 1). The effects of the deletion in the SFFV *env* gene are depicted in Fig. 2, where the structure of gp52 is compared with that of a typical retroviral envelope protein.

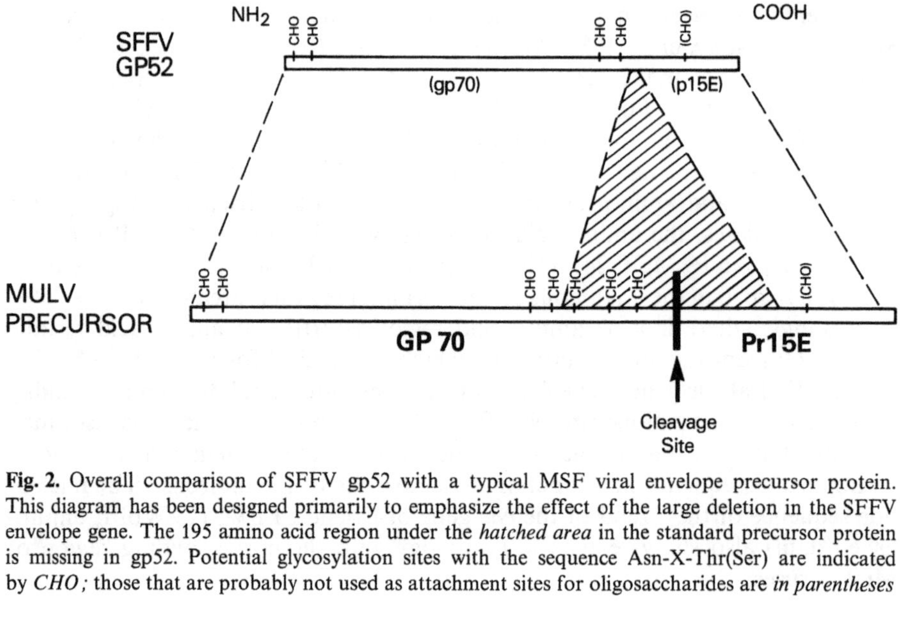

Fig. 2. Overall comparison of SFFV gp52 with a typical MSF viral envelope precursor protein. This diagram has been designed primarily to emphasize the effect of the large deletion in the SFFV envelope gene. The 195 amino acid region under the *hatched area* in the standard precursor protein is missing in gp52. Potential glycosylation sites with the sequence Asn-X-Thr(Ser) are indicated by *CHO*; those that are probably not used as attachment sites for oligosaccharides are *in parentheses*

The location of the SFFV envelope deletion was only recently determined by nucleotide sequence analysis (AMANUMA et al. 1983; WOLFF et al. 1983; CLARK and MAK 1983). Although heteroduplexes between SFFV and F-MuLV had previously indicated that a large deletion existed in the envelope gene of SFFV (BOSSELMAN et al. 1980; M. Gonda, personal communication), it was thought that gp52 was simply a truncated gp70. One reason for this assumption was the inability to precipitate gp52 with antiserum prepared to p15E (SCHULTZ et al. 1980). In support of the nucleotide sequence data, we now have evidence at the protein level that p15E is translated as a part of gp52. This evidence comes from a comparison of tryptic peptide fingerprints of gp52 with fingerprints of p15Es from F-MuLV and Friend MCF virus (WOLFF et al. 1984).

The carboxyl-terminal domain of p15E that is translated as a part of gp52 is that portion which contains the p15E transmembrane sequence of nonpolar amino acids (LENZ et al. 1982; PINTER and HONNEN 1983; KOCH et al. 1983). It is reasonable to assume that this nonpolar region also provides a transmembrane anchor for the SFFV envelope glycoprotein, which is clearly associated with membranes and is not secreted from the cell (RUSCETTI et al. 1979; DRESLER et al. 1979; LYLES and MCCONNELL 1981; SRINIVAS and COMPANS 1983). Consistent with this idea is the recent finding that, like p15E (A. Schultz, personal communication), the plasma membrane form of gp52 (gp65) can be labeled with [^{3}H]palmitic acid (SRINIVAS and COMPANS 1983). This and other fatty acids have been shown to be incorporated post-translationally into a wide variety of membrane glycoproteins and may serve to increase the affinity of glycoproteins for membranes (SCHMIDT and SCHLESINGER 1980; SCHMIDT 1982).

6.2 Carboxyl Terminus

Striking features within the carboxyl terminus of gp52 became evident with nucleotide sequencing of the SFFV *env* gene (AMANUMA et al. 1983; WOLFF et al. 1983; CLARK and MAK 1983). The p15E-related domain of gp52 is different from the prototype p15E in that its C-terminus is shorter and has a unique set of amino acids. These effects are due to a single base pair insertion that shifts the reading frame. Such an insertion has been found in every SFFV *env* gene sequenced to date, including the *env* genes of three Friend $SFFV_P$s, Friend $SFFA_A$ (L. Wolff unpublished data), and Rauscher SFFV (R. Bestwick and D. Kabat, unpublished data), the latter of which was clearly derived independently of the others. Interestingly, neither the exact position of the insertion nor the particular base inserted is the same for each isolate of SFFV sequenced, although the effects of the insertion in every case are to shorten the reading frame by 99 base pairs (33 amino acids) and produce a unique five to six amino acid sequence. Since all isolates of SFFV have been independently passaged for many years in vivo and at least one isolate clearly has an independent origin, the remarkable similarity of the novel carboxyl terminal ends of these SFFV envelope proteins suggests that this genetic modification may be crucial for generating a pathogenic SFFV.

One effect of the single base pair insertion is the loss of the R peptide from the carboxyl end of gp52. The role of this protein in the standard retrovirus is not known, but it has been suggested that it may play a role in virus maturation or act as a signal for specific protein compartmentalization (GREEN et al. 1981). It is possible that the loss of the R peptide could account solely for the pathogenic effects of the virus. However, the nucleotide sequence indicates that insertions in the envelope gene of SFFV could have occurred anywhere downstream within a 36-base pair region and still produce a termination codon prior to the R peptide. The fact that the insertions all occurred in the same small (approximately 3 base pair) area for all isolates of SFFV suggests that the specific length of the C-terminus and/or the unique set of amino acids produced may also be important for pathogenicity.

Another genetic modification at the 3′ end of the *env* gene observed for some strains of SFFV (AMANUMA et al. 1983; WOLFF et al. 1983; CLARK and MAK 1983) is a duplication of 6 base pairs. This results in the addition of two leucine residues to the membrane-spanning region of gp52.

6.3 Glycosylation

The glycosylation pattern of the envelope gene product of SFFV is unique compared with that of all other retroviral envelope glycoproteins. Gp52 has fewer glycosylation attachment sites, and the sugar processing that occurs at these sites is for the most part incomplete and heterogeneous. Studies by several investigators have shown that the predominant form of gp52 that accumulates intracellularly contains immature oligosaccharides of the high mannose-containing type (RUSCETTI et al. 1979; DRESLER et al. 1979). This form of the protein has the same glycopeptide profile as the envelope precursors of F-MuLV and Friend MCF virus (KEMP et al. 1981), consisting of two major size classes of 2200 and 1500 daltons (SRINIVAS and COMPANS 1983). Gp52 differs, however, from these envelope precursors in the number of oligosaccharide attachment sites. Whereas other retroviral envelope precursors have seven or eight attachment sites (BOSSELMAN et al. 1982; KOCH et al. 1983), gp52 has only four (SRINIVAS and COMPANS 1983). Three of the missing sites can be accounted for by the large deletion in the envelope gene (see Figs. 1, 2).

Although the majority of the gap52 molecules accumulate intracellularly as if blocked from passage into the Golgi apparatus where the final sugar processing occurs, a small proportion of gp52 molecules passes through the entire pathway used for processing of glycoproteins. The processed protein, which incorporates a variety of complex sugars such as galactose, fucose, and glucosamine, has a molecular weight of 65000 daltons and can be detected on the cell surface (RUSCETTI et al. 1979; GILLIS et al. 1979; RUTA and KABAT 1980). The glycosylation pattern of the 65K SFFV envelope protein indicates that it is more closely related to Friend MCF gp70 than to F-MuLV gp70 (SRINIVAS and COMPANS 1983). Like MCF gp70 (KEMP et al. 1980), SFFV gp65 lacks one of the two complex glycopeptides (the 5100-dalton G_1 glycopeptide) found in ecotropic viral gp70s (KEMP et al. 1979) but retains the 2900-dalton

G_2 glycopeptide (SRINIVAS and COMPANS 1983). SFFV gp65 differs from both the ecotropic and MCF viral gp70s in that it contains an additional glycopeptide of 2500 daltons (SRINIVAS and COMPANS 1983).

There have been speculations on the cause for the inefficient processing of the SFFV envelope glycoprotein. RUTA and co-workers (1982) have suggested that the unique structure of gp52 allows for spontaneous denaturation or partial unfolding of the protein while en route from the endoplasmic reticulum to the Golgi apparatus, resulting in heterogeneous cessation of processing. It has also been proposed (SRINIVAS and COMPANS 1983) that the *env* gene deletion in SFFV has resulted in the loss of a specific recognition signal involved in packaging glycoproteins within transport vesicles (ROTHMAN and FINE 1980; BLOBEL 1980).

Whether or not the aberrant processing of gp52 is critical to disease induction by the protein has not been determined. It is certainly possible that the intracellular accumulation of this protein within the endoplasmic reticulum could interfere with normal cellular functions.

7 Molecular Basis for the $SFFV_P$ and $SFFV_A$ Phenotypes

Several studies have compared the anemia and polycythemia-inducing variants of SFFV at both the genetic and protein levels. The purpose of these experiments has been to localize the determinant(s) in the virus that might be responsible for the two different phenotypes of the disease.

As mentioned above, both $SFFV_A$ and $SFFV_P$ encode 52000- to 55000-dalton MCF gp70-related envelope glycoproteins. Although differences can be found in the tryptic peptides when the envelope proteins of $SFFV_A$ and $SFFV_P$ are compared (MACDONALD et al. 1980a), it is not known whether such differences relate to differences in biological activity since differences in peptide composition of the envelope proteins exist among different isolates of $SFFV_P$ (YOSHIDA and YOSHIKURA 1981; S. Ruscetti, unpublished data). One difference that appears to exist between the envelope glycoproteins of polycythemia- and anemia-inducing isolates of SFFV is in post-translational processing of the protein (RUSCETTI et al. 1981b). While a small percentage of the gp52 encoded by all $SFFV_P$ isolates is processed by the addition of complex sugars to a 65K form that appears on the cell surface, both Friend $SFFV_A$ (RUSCETTI et al. 1981b) and the anemia-inducing Rauscher SFFV (RUTA and KABAT 1980) are deficient in this process and little if any gp65 can be detected on the cell surface.

When the RNAs of $SFFV_P$ and $SFFV_A$ were compared by oligonucleotide fingerprinting (EVANS et al. 1980), it was found that $SFFV_A$ contained several T_1-oligonucleotides that were not present in $SFFV_P$, F-MuLV, or MCF viruses. These $SFFV_A$-specific oligonucleotides were situated in two locations: at the very 5′ end of the genome and in a central region of the genome that may include the 3′ end of the *pol* gene and/or the 5′ end of the *env* gene. None of these oligonucleotides, however, appear to be involved in determining the distinct biological phenotype of $SFFV_A$ (see p. 36).

Studies with subgenomic fragments of molecularly cloned $SFFV_A$ have localized the area determining the phenotype of the virus to the *env* gene (Kaminchik et al. 1982). In more recent experiments (J. Kaminchik, E. Scolnick, and D. Hankins, submitted for publication), recombinants between the envelope genes of molecularly cloned $SFFV_A$ and $SFFV_P$ have been engineered. From an analysis of the biological phenotypes of the recombinants in vivo (polycythemia versus anemia) and in vitro (development of hemoglobinized versus underhemoglobinized vBFU-E in the absence of Epo), it was concluded that the genetic information required for expression of one or the other phenotype resided in the 3′ half of the envelope gene.

8 Conclusions and Speculations

The data clearly indicate that the envelope gene of the spleen focus-forming virus plays an essential role in the initiation of the erythroproliferative disease induced by SFFV. The other genes in SFFV appear either to be nonessential or can be easily substituted by those contained in any of a number of MuLVs. However, mutations in the envelope gene of SFFV clearly destroy the pathogenicity of the virus, and no other MuLVs can substitute for it. Although the LTR region of the retroviral genome has been implicated in diseases caused by some leukemia viruses (Neel et al. 1981; Payne et al. 1981), there is no evidence that the LTR plays a similar role in the acute erythroproliferative disease induced by SFFV. Mutants of SFFV that delete most of the LTR region are still pathogenic. On the other hand, it is quite possible that the SFFV LTR could play a role in the late, malignant diseases induced by SFFV (see below).

The envelope genes and gene products of all SFFVs studied to date are remarkably similar and share a number of unusual features which may be responsible for the increased pathogenicity of these viruses. One feature is a close homology with the envelope proteins contained in MCF viruses, which are clearly the pathogenic intermediates involved in leukemias induced by MuLVs. Both SFFV and MCF viruses have acquired *env* gene sequences normally present in the mouse DNA. Perhaps the proteins encoded by these endogenous retroviral envelope genes play specific roles in normal growth and differentiation of hematopoietic cells. If so, increased expression of such endogenous genes by recombination with MuLVs, resulting in MCF viruses and SFFV, could have profound effects on a variety of hematopoietic pathways. In support of such an idea is a recent observation by Portis et al. (1983) that a a monoclonal antibody, produced during graft-versus-host reaction, that appears to be directed toward an erythroid differentiation antigen also reacts with the envelope protein encoded by an endogenous retrovirus.

Although all isolates of SFFV are highly related to MCF viruses, there are certain features of their *env* genes and products that are unique to SFFVs and conserved in every strain. These particular features may be responsible for the unique manifestation of erythroid disease induced by SFFV. All isolates of SFFV analyzed to date have a centrally located deletion in the open reading frame of the *env* gene. This results in a gene product, considerably smaller than the standard retroviral envelope precursor, which has covalently linked terminal portions of gp70 and p15E. In addition, all isolates of SFFV have been shown to have a single base pair insertion which results in a change in reading frame, translation of a short unique sequence, and premature termination. The observation that all isolates of SFFV have similar deletions and insertions in their envelope genes raises the question as to whether these genetic features already existed in the endogenous viral envelope gene that recombined with ecotropic virus in the generation of SFFV or whether they occurred subsequent to recombination.

In order to determine the individual features of the SFFV *env* gene which are required for pathogenicity, more precise manipulation of the gene, by recombination with other MuLVs and by in vitro mutagenesis, will be necessary. Along these lines, S.M. Anderson (personal communication) has prepared recombinants, using the *Eco*R1 site, between various parts of the SFFV *env* gene and the *env* genes of either amphotropic or AKR MCF viruses. Preliminary studies indicate that recombinants containing the 3′ half of the SFFV *env* gene and the 5′ half of the amphotropic or MCF *env* gene were pathogenic but recombinants containing the 5′ half of the SFFV *env* gene and the 3′ half of the amphotropic *env* gene were not. These data suggest that the unique features contained in the 3′ half of the SFFV *env* gene are crucial for pathogenicity and cannot be substituted for by other retroviral *env* gene sequences. On the other hand, the 5′ half of the SFFV *env* gene, which contains the MCF-related sequences, can be substituted for by other retroviral *env* gene sequences. It should be emphasized, however, that both of the two particular viruses used in this reconstruction experiment, amphotropic and AKR MCF virus, are dualtropic and contain 5′ *env* gene sequences that are more related to those contained in SFFV than F-MuLV. Thus, 5′ *env* gene sequences from any dualtropic virus may be able to substitute for 5′ SFFV *env* gene sequences. Additional studies are needed to determine whether these 5′ *env* gene sequences can derive from retroviruses other than those with a dualtropic host range.

It is interesting to speculate on how the SFFV envelope glycoprotein could interfere with erythropoiesis. Since the polycythemia- and anemia-inducing variants of SFFV encode such similar envelope glycoproteins, it is likely that they alter erythropoiesis by a similar mechanism. A common mechanism that would best fit the in vitro data would be one based on SFFV making an erythroid precursor more sensitive to Epo than an uninfected cell, with $SFFV_P$ making the cell the most sensitive and $SFFV_A$ somewhat less. This could be accomplished in a number of ways. For example, the SFFV envelope protein may have the same effect on the cell as Epo when it is expressed in an Epo-responsive cell. Alternatively, the SFFV envelope protein could cross-react with the receptor for Epo, enabling the cell to bind more Epo, thus reducing the concentration of exogenous hormone required for differentiation in vitro. Since $SFFV_P$ and $SFFV_A$ differ in their ability to encode the plasma membrane form of the envelope protein, then $SFFV_P$ would be more effective than $SFFV_A$ in modifying erythroid differentiation if Epo interacts with its receptor at the cell surface. These and other models can be tested as purified proteins and assays to detect Epo receptors become available. It should be mentioned that Friend MCF virus, which encodes an envelope glycoprotein that is closely related to, yet distinct from, the SFFV envelope protein, is also thought to have effects on erythropoiesis. Unlike SFFV-infected cells, cells infected with Friend MCF virus do not have increased sensitivity to Epo and, in fact, they appear to be blocked at an early stage of erythroid differentiation (MacDonald et al. 1980b; Niho et al. 1982). It will be interesting to determine whether differences in the structures of SFFV and Friend MCF envelope proteins, in their post-translational processing and/or in their subcellular locations are responsible for their distinct effects on particular stages of erythroid differentiation.

While it is clear that the envelope gene of SFFV is involved in the acute pathogenic effects of the virus on erythropoiesis, it is not known what is necessary to induce the late, malignant effects of the Friend virus complex. Molecularly cloned SFFV preparations have been used for some of these studies (Yamamoto et al. 1981), which rules out a contaminating oncogenic virus in the inocula being responsible for the establishment of malignant cells. It is possible that SFFV is responsible for the downstream promotion or rearrangement of a cellular oncogene that is now expressed in elevated levels in these cells. Although no evidence to date supports that theory, chromosomal rearrangements have been detected in a number of erythroleukemia cell lines from SFFV-infected mice (Ostertag et al. 1972). Alternatively, SFFV infection, by causing increased proliferation of erythroid precursor cells, may increase the probability of a spontaneous mutational event resulting in malignancy. Consistent with this possibility is a recent study (Shibuya and Mak 1983) showing that the erythroleukemia lines isolated late in the disease appear to be at the same level of differentiation as the proliferating cells in the corresponding early stages. It will be interesting to determine whether some of the mutants of SFFV discussed above that do not induce acute erythroid hyperplasia are capable of inducing the late malignant diseases. Finally, it is possible that SFFV plays no role in these late transplantable diseases induced by the Friend virus complex. Transplantable cell lines have been developed only from mice infected with the original Friend virus complex and it is now clear that the helper virus in the complex, F-MuLV, can itself induce a variety of late malignant diseases in adult mice (Chesebro et al. 1983; P. Tambourin, personal communication). In order to determine the role of SFFV in the development of these malignant lines, it will be necessary to develop lines from mice infected with SFFV rescued with a nonpathogenic helper virus or with SFFV alone.

Acknowledgments. We would like to thank Steven Anderson, Richard Bestwick, Matthew Gonda, David Hankins, David Kabat, Joseph Kaminchik, and Ranga Srinivas for sharing their unpublished data with us. In addition, we are very grateful to Frank Ruscetti and Pierre Tambourin for critically reading this manuscript.

References

Amanuma H, Akiko K, Obata M, Sagata N, Ikawa Y (1983) Complete nucleotide sequence of the gene for the specific glycoprotein (gp55) of Friend spleen focus-forming virus. Proc Natl Acad Sci USA 80:3913–3917

Anand R, Lilly F, Ruscetti S (1981) Viral protein expression in producer and nonproducer clones of Friend erythroleukemia cell lines. J Virol 37:654–660

Axelrad AA, Steeves RA (1964) Assay for Friend leukemia virus: rapid quantitative method based on enumeration of macroscopic spleen foci in mice. Virology 24:513–518

Barbacid M, Troxler DH, Scolnick EM, Aaronson SA (1978) Analysis of translational products of Friend strain of spleen focus-forming virus. J Virol 27:826–830

Bernstein A, Mak TW, Stephenson JR (1977) The Friend virus genome: evidence for the stable association of MuLV sequences and sequences involved in erythroleukemic transformation. Cell 12:287–294

Bestwick R, Ruta M, Kiessling A, Faust C, Linemeyer D, Scolnick E, Kabat D (1983) Genetic structure of Rauscher spleen focus-forming virus. J Virol 45:1217–1222

Bilello JA, Colletta G, Warnecke G, Koch G, Frisby D, Pragnell IB, Ostertag W (1980) Analysis of the expression of spleen focus-forming virus (SFFV)-related RNA and gp55, a Friend and Rauscher virus-specific protein. Virology 107:331–344

Blobel G (1980) Intracellular protein topogenesis. Proc Natl Acad Sci USA 77:1496–1500

Bosselman RA, Van Griensven LJLD, Vogt M, Verma I (1980) Genome organization of retroviruses IX. Analysis of the genomes of Friend spleen focus-forming (F-SFFV) and helper murine leukemia viruses by heteroduplex-formation. Virology 102:234–239

Bosselman RA, van Straaten F, Van Beveren C, Verma IM, Vogt M (1982) Analysis of the *env* gene of a molecularly cloned and biologically active Moloney mink cell focus-forming proviral DNA. J Virol 44:19–31

Chattopadhyay SK, Cloyd MW, Linemeyer DL, Lander MR, Rands E, Lowy DR (1982) Cellular origin and role of mink cell focus-forming viruses in murine thymic lymphomas. Nature 295:25–31

Chesebro B, Portis JL, Wehrly K, Nishio J (1983) Effect of murine host genotype on MCF virus expression, latency and leukemia cell type of leukemias induced by Friend murine leukemia helper virus. Virology 128:221–233

Clark SP, Mak TW (1982) Nucleotide sequences of the murine retrovirus Friend $SFFV_P$ long terminal repeats: identification of a structure with extensive dyad symmetry 5′ to the TATA box. Nucleic Acids Res 10:3315–3330

Clark SP, Mak TW (1983) Complete nucleotide sequences of an infectious clone of Friend spleen focus-forming provirus: gp55 is an envelope fusion glycoprotein. Proc Natl Acad Sci USA 80:5037–5041

Dexter TM, Scott D, Teich NM (1977) Infection of bone marrow cells in vitro with FLV: effects on stem cell proliferation, differentiation and leukemogenic capacity. Cell 12:355–364

Dexter TM, Allen TD, Scott D, Teich NM (1979) Isolation and characterization of a bipotential haematopoietic cell line. Nature 227:471–474

Dexter TM, Allen TD, Testa NG, Scolnick E (1981) Friend disease in vitro. J Exp Med 154:594–608

Dresler S, Ruta M, Murray MJ, Kabat D (1979) Glycoprotein encoded by the Friend spleen focus-forming virus. J Virol 30:564–575

Evans LH, Duesberg PH, Troxler DH, Scolnick EM (1979) Spleen focus-forming Friend virus: identification of genomic RNA and its relationship to helper virus RNA. J Virol 31:133–146

Evans L, Nunn M, Duesberg PH, Troxler D, Scolnick E (1980) RNAs of defective and nondefective components of Friend anemia and polycythemia virus strains identified and compared. Cold Spring Harbor Symp Quant Biol 44:823–835

Fagg B, Ostertag W (1982) Friend erythroleukemia virus complex: role of viral components in modifying erythroid differentiation in mice. J Natl Cancer Inst 68:457–460

Famulari NG (1983) Murine leukemia viruses with recombinant *env* genes: a discussion of their role in leukemogenesis. In: Vogt PK, Koprowski H (eds) Retroviruses 1. Springer, Berlin Heidelberg New York, pp 76–108 (Current topics in microbiology and immunology, vol 103)

Fredrickson T, Tambourin P, Wendling F, Jasmin C, Smajda F (1975) Target cell of the polycythemia-inducing Friend virus: studies with myleran. J Natl Cancer Inst 55:443–446

Friend C (1957) Cell-free transmission in adult Swiss mice of a disease having the character of a leukemia. J Exp Med 105:307–318
Friend C, Haddad JR (1960) Tumor formation with transplants of spleen or liver from mice with virus-induced leukemia. J Natl Cancer Inst 25:1279–1289
Friend C, Sher W, Holland JG, Sato T (1971) Hemoglobin synthesis in murine virus-induced leukemia cells in vitro: stimulation of erythroid differentiation by dimethylsulfoxide. Proc Natl Acad Sci USA 68:378–382
Gillis S, Ruscetti SK, Gillis AE, Troxler DH, Scolnick EM, Smith KA (1979) The spleen focus-forming virus (SFFV)-specific neoantigen shares cross-reactive determinants with mink cell focus-inducing (MCF) virus gp70. Virology 96:421–428
Green N, Shinnick TM, Owen W, Ponticelli A, Sutcliff JG, Lerner RA (1981) Sequence-specific antibodies show that maturation of Moloney leukemia virus envelope polyprotein involves removal of a COOH-terminal peptide. Proc Natl Acad Sci USA 78:6023–6027
Greenberger JS, Davisson PB, Gans PJ, Moloney WC (1979) In vitro induction of continuous acute promyelocytic leukemia cell lines by Friend or Abelson murine virus. Blood 53:987–1001
Gregory CJ, McCulloch EA, Till JE (1973) Erythropoietic progenitors capable of colony formation in culture: state of differentiation. J Cell Physiol 81:411–420
Hankins WD, Krantz SB (1975) In vitro expression of erythroid differentiation induced by Friend polycythemia virus. Nature 253:731–732
Hankins WD, Kratz SB (1980) Helper virus is not required for in vitro erythroid transformation of hematopoietic cells by Friend virus. Proc Natl Acad Sci USA 77:5287–5291
Hankins WD, Troxler D (1980) Polycythemia- and anemia-inducing erythroleukemia viruses exhibit differential erythroid transforming effects in vitro. Cell 22:693–699
Hankins WD, Kost TA, Khoury MJ, Krantz SB (1978) Erythroid bursts produced by Friend leukemia virus in vitro. Nature 276:506–508
Holland CA, Wozney J, Hopkins N (1983) Nucleotide sequence of the gp70 gene of murine retrovirus MCF 247. J Virol 47:413–420
Horoszewicz JS, Leong SS, Carter WA (1975) Friend leukemia: rapid development of erythropoietin-independent hematopoietic precursors. J Natl Cancer Inst 54:265–267
Ikawa Y, Yoshida M, Yoshikura H (1978) Identification of proteins specific to Friend strain of spleen focus forming virus (SFFV). Proc Jpn Acad 54:651–656
Ikawa Y, Kobayashi Y, Obinata M, Harada F, Hino S, Yoshikura H (1980) RNA sequences and proteins specific to Friend strain of spleen focus-forming virus. Cold Spring Harbor Symp Quant Biol 44:875–885
Iscove NN (1978) Erythropoietin-independent stimulation of early erythropoiesis in adult bone marrow cultures by conditioned medium from lectin-stimulated mouse spleen cells. In: Golde DW, Cline MJ, Metcalf D, Fox CF (eds) Hematopoietic cell differentiation. Academic, New York, pp 37–52
Johnson CS, Marcelletti J, Longley C, Furmanski P (1982) Inhibition of normal erythropoiesis in mice with Friend virus induced erythroleukemia. Exp Hematol 10:743–753
Kaminchik J, Hankins WD, Ruscetti SK, Linemeyer DL, Scolnick EM (1982) Molecular cloning of biologically active proviral DNA of the anemia-inducing strain of spleen focus-forming virus. J Virol 44:922–931
Kemp MC, Basak S, Compans RW (1979) Glycopeptides of murine leukemia virus. I. Comparison of two ecotropic viruses. J Virol 31:1–7
Kemp MC, Famulari NG, O'Donnell PV, Compans RW (1980) Glycopeptides of murine leukemia virus II. Comparison of xenotropic and dualtropic viruses. J Virol 34:154–161
Kemp MC, Famulari NG, Compans RW (1981) Glycopeptides of murine leukemia virus III. Glycosylation of *env* precursor glycoproteins. J Virol 39:463–470
Klein G (ed) (1982) Advances in viral oncology, vol 2. Raven, New York
Koch W, Hunsmann G, Friedrich R (1983) Nucleotide sequence of the envelope gene of Friend murine leukemia virus. J Virol 45:1–9
Koch W, Zimmermann W, Oliff A, Friedrich R (1984) Molecular analysis of the envelope gene and long terminal repeat of Friend mink cell focus-inducing virus: Implications for the functions of these sequences. J Virol 49:828–840
Kost TA, Koury MJ, Hankins WD, Krantz SB (1979) Target cells for Friend virus-induced erythroid bursts in vitro. Cell 18:145–152

Kost TA, Hankins WD, Krantz SB (1980) Combined effect of Friend polycythemia virus and erythropoietin on erythroid burst formation in vitro. Exp Hematol 8:248–258

Kost TA, Koury MJ, Krantz SB (1981) Mature erythroid burst forming units are target cells for Friend virus-induced erythroid bursts. Virology 108:309–317

Koury MJ, Bondurant MC, Duncan DT, Krantz SB, Hankins WD (1982) Specific differentiation events induced by erythropoietin in cells infected in vitro with the anemia strain of Friend virus. Proc Natl Acad Sci USA 79:635–639

Langdon WY, Hoffman PM, Silver JE, Buckler CE, Hartley JW, Ruscetti SK, Morse HC (1983a) Identification of a spleen focus-forming virus in erythroleukemic mice infected with a wild-mouse ecotropic murine leukemia virus. J Virol 46:230–238

Langdon WY, Ruscetti SK, Silver JE, Hankins WD, Buckler CE, Morse HC (1983b) Cas spleen focus-forming virus. II. Further biological and biochemical characterization. J Virol 47:329–336

Lenz J, Crowther R, Straceski A, Haseltine W (1982) Nucleotide sequence of the Akv *env* gene. J Virol 42:519–529

Levy SB, Blankstein LA, Vinton EC, Chamber TJ (1979) In: Ikawa Y, Odaka T (eds) Oncogenic viruses and host cell genes. Academic, New York, pp 409–428

Liao S, Axelrad AA (1975) Erythropoietin-independent erythroid colony formation in vitro by hemopoietic cells of mice infected with Friend virus. Int J Cancer 15:467–482

Linemeyer DL, Ruscetti SK, Menke JG, Scolnick EM (1980) Recovery of biologically active spleen focus-forming virus from molecularly cloned spleen focus-forming virus-pB322 circular DNA by cotransfection with infectious type C retroviral DNA. J Virol 35:710–721

Linemeyer DL, Ruscetti SK, Scolnick EM, Evans LH, Duesberg PH (1981) Biological activity of the spleen focus-forming virus is encoded by a molecularly cloned subgenomic fragment of spleen focus-forming virus DNA. Proc Natl Acad Sci USA 78:1401–1405

Linemeyer DL, Menke JG, Ruscetti SK, Evans LH, Scolnick EM (1982) Envelope gene sequences which encode the gp52 protein of spleen focus-forming virus are required for the induction of erythroid cell proliferation. J Virol 43:223–233

Lyles DS, McConnell KA (1981) Subcellular localization of the *env*-related glycoproteins in Friend erythroleukemia cells. J Virol 39:263–272

MacDonald ME, Reynolds FH, Van de Ven WJM, Stephenson JR, Mak TW, Bernstein A (1980a) Anemia- and polycythemia-inducing isolates of Friend spleen focus-forming virus. Biological and molecular evidence for two distinct viral genomes. J Exp Med 151:1447–1492

MacDonald ME, Mak TW, Bernstein A (1980b) Erythroleukemia induction by replication-competent type-C viruses cloned from the anemia- and polycythemia-inducing isolates of Friend leukemia virus. J Exp Med 151:1493–1503

MacDonald ME, Johnson GR, Bernstein A (1981) Different pseudotypes of Friend spleen focus-forming virus induce polycythemia and erythropoietin-independent colony formation in serum-free medium. Virology 110:231–236

Machida CA, Bestwick RK, Kabat D (1984) Reduced leukemogenicity caused by mutations in the membrane glycoprotein gene of Rauscher spleen focus-forming virus. J Virol 49:394–402

Mager D, Mak TW, Bernstein A (1980) Friend leukemia virus-transformed cells, unlike normal stem cells, form spleen colonies in Sl/Sl^d mice. Nature 288:592–594

Mager DL, Mak TW, Bernstein A (1981) Quantitative colony method for tumorigenic cells transformed by two distinct strains of Friend leukemia virus. Proc Natl Acad Sci USA 78:1703–1707

Metcalf D, Furth J, Buffett RF (1959) Pathogenesis of mouse leukemia caused by Friend virus. Cancer Res 19:52–60

Mirand EA (1967) Virus-induced erythropoiesis in hypertransfused polycythemic mice. Science 156:832–833

Mirand EA, Prentice TC, Hoffmann JG (1961) Effect of Friend virus in Swiss and DBA/1 mice on ^{59}Fe uptake. Proc Soc Exp Biol Med 106:423–426

Mirand EA, Steeves RA, Ayila L, Grace JT (1968a) Spleen focus formation by polycythemic strains of Friend leukemia virus. Proc Soc Exp Biol Med 127:900–904

Mirand EA, Steeves RA, Lange RD, Grace JT (1968b) Virus-induced polycythemia in mice: erythropoiesis without erythropoietin. Proc Soc Exp Biol Med 128:844–849

Neel BG, Hayward WS, Robinson HL, Fang J, Astrin S (1981) Avian leukosis virus-induced tumors have common proviral integration sites and synthesize discrete new RNAs: oncogenesis by promoter insertion. Cell 23:323–334

Niho Y, Shibuya T, Mak TW (1982) Modulation of erythropoiesis by the helper-independent Friend leukemia virus F-MuLV. J Exp Med 156:146–158

Opitz V, Seidel HJ (1978) Studies on the target cell for the Friend virus (FV-P) strain using the CFU-E technique. Blut 37:183–192

Ostertag W, Melderis H, Steinheider G, Kluge N, Dube S (1972) Synthesis of mouse hemoglobin and globin mRNA in leukaemic cell cultures. Nature 239:231–234

Payne GS, Courtneidge SA, Crittenden LB, Fadly AM, Bishop JM, Varmus HE (1981) Analysis of avian leukosis virus DNA and RNA in bursal tumors: viral gene expression is not required for maintenance of the tumor state. Cell 23:311–322

Peschle C, Colletta G, Covelli A, Ciccariello R, Migliaccio G, Rossi GB (1981) The erythropoietic component of Friend virus erythroleukemias: role of erythropoietic hormones and SFFV genome. In: Rovoltella R, Pontieri G, Rovera G, Basilico C, Gallo RC, Subak-Sharpe J (eds) Expression of differentiated functions in cancer cells. Raven, New York, pp 311–321

Pinter A, Honnen WJ (1983) Topography of murine leukemia virus envelope protein: characterization of transmembrane components. J Virol 46:1056–1060

Pluznik DH, Sachs L (1964) Quantitation of a murine leukemia virus with a spleen colony assay. J Natl Cancer Inst 35:535–564

Portis JL, Britt WJ, Evans LH, Chesebro B (1983) An antibody produced during graft-versus host reaction recognizes an erythroid differentiation antigen. Leukemia Rev 1:211–212

Racevskis J, Koch G (1978) Synthesis and processing of viral proteins in Friend erythroleukemia cell lines. Virology 87:354–365

Rauscher FJ (1962) A virus-induced disease of mice characterized by erythrocytopoiesis and lymphoid leukemia. J Natl Cancer Inst 29:515–543

Rein A (1982) Interference grouping of murine leukemia viruses: a distinct receptor for the MCF-recombinant viruses on mouse cells. Virology 120:251–257

Rossi GB, Peschle C (1980) Enhanced proliferation and migration of BFU-E and erythropoietin-independence of CFU-E expression in FLV-infected mice: comparative studies on anemic and polycythemic strain. In: Rossi GB (ed) In vivo and in vitro erythropoiesis: the Friend system. Elsevier, Amsterdam, pp 139–149

Rothman JE, Fine RE (1980) Coated vesicles transport newly synthesized membrane glycoproteins from endoplasmic reticulum to plasma membrane in two successive stages. Proc Natl Acad Sci USA 77:780–784

Ruscetti SK, Scolnick EM (1983) Expression of a transformation-related protein (p53) in the malignant stage of Friend virus-induced diseases. J Virol 46:1022–1026

Ruscetti S, Linemeyer D, Feild J, Troxler D, Scolnick E (1978) Type-specific radioimmunoassays for the gp70s of mink cell focus-inducing murine leukemia viruses: expression of a cross-reacting antigen in cells infected with the Friend strain of the spleen focus-forming virus. J Exp Med 148:654–663

Ruscetti S, Linemeyer D, Feild J, Troxler D, Scolnick EM (1979) Characterization of a protein found in cells infected with the spleen focus-forming virus that shares immunological cross-reactivity with the gp70 found in mink cell focus-inducing virus particles. J Virol 30:787–798

Ruscetti S, Troxler D, Linemeyer D, Scolnick E (1980) Three laboratory strains of spleen focus-forming virus: comparison of their genomes and translational products. J Virol 33:140–151

Ruscetti S, Davis L, Feild J, Oliff A (1981a) Friend murine leukemia virus-induced leukemia is associated with the formation of mink cell focus-inducing viruses and is blocked in mice expressing endogenous mink cell focus-inducing xenotropic viral envelope genes. J Exp Med 154:907–920

Ruscetti SK, Feild JA, Scolnick EM (1981b) Polycythaemia- and anaemia-inducing strains of spleen focus-forming virus differ in post-translational processing of envelope-related glycoproteins. Nature 294:663–665

Ruta M, Kabat D (1980) Plasma membrane glycoproteins encoded by cloned Rauscher and Friend spleen focus-forming viruses. J Virol 35:844–853

Ruta M, Clarke S, Boswell B, Kabat D (1982) Heterogeneous metabolism and subcellular localization of a potentially leukemogenic membrane glycoprotein encoded by Friend erythroleukemia virus. J Biol Chem 257:126–134

Ruta M, Bestwick R, Machida C, Kabat D (1983) Loss of leukemogenicity caused by mutations in the membrane glycoprotein structural gene of Friend spleen focus-forming virus. Proc Natl Acad Sci USA 80:4704–4708

Sassa S, Takaku F, Nakao K (1968) Regulation of erythropoiesis in the Friend leukemia mouse. Blood 31:758–765

Schmidt M (1982) Acylation of viral spike glycoproteins: a feature of enveloped RNA viruses. Virology 116:327–338

Schmidt MFG, Schlesinger MJ (1980) Relationship of fatty acid attachment to the translation and maturation of vesicular stomatitis and Sindbis virus membrane glycoproteins. J Biol Chem 255:3334–3339

Schultz AM, Ruscetti SK, Scolnick EM, Oroszlan S (1980) The *env*-gene of the spleen focus-forming virus lacks expression of p15(E) determinants. Virology 107:537–542

Shibuya T, Mak TW (1983) Isolation and induction of erythroleukemia cell lines with properties of erythroid progenitor burst-forming cell (BFU-E) and erythroid precursor cell (CFU-E). Proc Natl Acad Sci USA 80:3721–3725

Srinivas RV, Compans RW (1983) Glycosylation and intracellular transport of spleen focus-forming virus glycoproteins. Virology 125:274–286

Srinivas RV, Compans RW (1983) Membrane association and defective transport of spleen focus-forming virus glycoproteins. J Biol Chem 258:14718–14724

Steeves RA, Mirand EA, Thomson S, Avila L (1969) Enhancement of spleen focus formation and virus replication in Friend virus-infected mice. Cancer Res 29:1111–1116

Steinheider G, Seidel HJ, Kreja L (1979) Comparison of the biological effects of anemia inducing and polycythemia inducing Friend virus complex. Experientia 35:1173–1175

Tambourin P, Wendling F (1971) Malignant transformation and erythroid differentiation by polycythaemia-inducing Friend virus. Nature 234:230–233

Tambourin P, Wendling F (1975) Target cell for oncogenic action of polycythaemia-inducing Friend virus. Nature 256:320–322

Tambourin PE, Wendling F, Jasmin C, Smadja-Joffe F (1979) The physiopathology of Friend leukemia. Leuk Res 3:117–129

Tambourin P, Wendling F, Moreau-Gachelin F, Charon M, Bucau-Varlet P (1980) Friend leukemia: a multi-step malignant disease. In: Rossi GB (ed) In vivo and in vitro erythropoiesis: the Friend system. Elsevier, Amsterdam, pp 127–138

Troxler DH, Scolnick EM (1978) Rapid leukemia induced by cloned Friend strain of replicating murine type-C virus. Virology 85:17–27

Troxler DH, Parks WP, Vass WC, Scolnick EM (1977a) Isolation of a fibroblast nonproducer cell line containing the Friend strain of the spleen focus-forming virus. Virology 76:602–615

Troxler DH, Boyars JK, Parks WP, Scolnick EM (1977b) Friend strain of spleen focus-forming virus: a recombinant between mouse type C ecotropic viral sequences and sequences related to xenotropic virus. J Virol 22:361–372

Troxler DH, Lowy D, Howk R, Young H, Scolnick EM (1977c) Friend strain of spleen focus-forming virus is a recombinant between ecotropic murine type C virus and the *env* gene region of xenotropic type C virus. Proc Natl Acad Sci USA 74:4671–4675

Troxler DH, Ruscetti SK, Linemeyer DL, Scolnick EM (1980) Helper-independent and replication-defective erythroblastosis-inducing viruses contained within anemia-inducing Friend virus complex (FV-A). Virology 102:28–45

Wagemaker G (1978) Induction of erythropoietin responsiveness in vitro. In: Golde DW, Cline MJ, Metcalf D, Fox CF (eds) Hematopoietic cell differentiation. Academic, New York, pp 109–118

Weiss R, Teich N, Varmus H, Coffin J (eds) (1982) RNA tumor viruses. Cold Spring Harbor Laboratory, Cold Spring Harbor

Wendling F, Moreau-Gachelin F, Tambourin P (1981) Emergence of tumorigenic cells during the course of Friend virus leukemias. Proc Natl Acad Sci USA 78:3614–3618

Wolff L, Koller R, Ruscetti S (1982) Monoclonal antibody to spleen focus-forming virus-encoded gp52 provides probe for the amino-terminal region of retroviral envelope proteins that confers dual tropism and xenotropism. J Virol 43:472–481

Wolff L, Scolnick E, Ruscetti S (1983) Envelope gene of the Friend spleen focus-forming virus: deletion and insertions in 3′ gp70/p15E-encoding region have resulted in unique features in the primary structure of its protein product. Proc Natl Acad Sci USA 80:4718–4722

Wolff L, Hubbert N, Ruscetti S (1984) Structural analysis of the spleen focus-forming virus envelope gene product. Virology 133:376–385

Yamamoto Y, Gamble CL, Clark SP, Joyner A, Shibuya T, MacDonald ME, Mager D, Bernstein A, Mak TW (1981) Clonal analysis of early and late stages of erythroleukemia induced by molecular clones of integrated spleen focus-forming virus. Proc Natl Acad Sci USA 78:6893–6897

Yoshida M, Yoshikura H (1980) Analysis of spleen focus-forming virus-specific RNA sequences coding for spleen focus-forming virus-specific glycoprotein with a molecular weight of 55,000 (gp55). J Virol 33:587–596

Yoshida M, Yoshikura H (1981) A glycoprotein specified by spleen focus-forming virus in three cell lines varies in molecular size and peptide composition, but retains both xenotropic and ecotropic MuLV antigenicity. J Gen Virol 52:183–188

The Genetics of Murine Leukemia Viruses

S.P. GOFF

1 Introduction . . . 45
2 The Wild-Type Viruses: Their Genes and Genomic Organization . . . 46
3 Mutants: Steps in the Life Cycle Mediated by Viral Genes . . . 49
3.1 Early Events: Adsorption and Uncoating of Virions . . . 49
3.2 Reverse Transcription . . . 51
3.3 Integration of the Viral DNA: Establishment of the Provirus . . . 53
3.4 Virion Assembly and Release . . . 55
3.5 Mutants Blocked in Packaging Viral RNA into Particles . . . 58
3.6 Leukemogenesis and Tissue Specificity . . . 59
3.7 Silent Mutations . . . 60
4 Genetic Events Mediated by Retroviruses . . . 61
4.1 Phenotypic Mixing and Complementation . . . 61
4.2 Genetic Recombination Between Viruses . . . 63
4.3 Recombination Between Virus and Host Genes . . . 64
4.4 Mutation of Cellular Genes . . . 65
References . . . 67

1 Introduction

The murine leukemia viruses are prototypical replication-competent retroviruses. These viruses were among the first known oncogenic agents in mammals and have been the subjects of intensive study for more than 3 decades (GROSS 1970). We now have a detailed understanding of the life cycle of these viruses and of some of the genetic effects they can have on the host cell; we are even beginning to discover how these viruses cause leukemia. Today, these viruses remain popular systems for the study of gene transfer and expression, site-specific recombination in mammalian cells, mutagenesis, and the processes of induction and progression of tumors. Most recently, their similarity to the human T-cell leukemia viruses (HTLVs) (POIESZ et al. 1982) and to sequences endogenous to the human genome (REPASKE et al. 1983) has stimulated interest in a still larger scientific community.

Like all retroviruses, the murine leukemia viruses undergo a complex series of genomic structural changes during their life cycle. The virions contain two copies of genomic RNA, which are approximately 8000 bases in length, capped at the 5′ end, and polyadenylated at the 3′ end. Soon after infection of sensitive

Department of Biochemistry and Institute for Cancer Research, Columbia University, College of Physicians and Surgeons, 630 West 168th Street, New York, NY 10032, USA

Current Topics in Microbiology and Immunology, Vol. 112

cells, a double-stranded linear DNA copy of the genomic RNA is synthesized; the reaction is not a simple one, involving DNA translocations (GILBOA et al. 1979) needed for the construction of special terminal sequences known as long terminal repeats (LTRs). The linear DNA product of the reaction is transported to the nucleus of the infected cell, and is circularized to yield two double-stranded covalently closed circular DNAs containing either one or two tandem copies of the LTR sequence. Finally, and most importantly for the viruses' genetic effects on the host, one or more forms of the viral DNA are integrated into the host genome by a mechanism closely related to the transposition of mobile genetic elements. The inserted DNA is stably maintained in the infected cell: effectively it becomes a gene, transcribed constitutively into mRNAs by the cell for the rest of the lifetime of the host. Such a stable producer cell continuously assembles viral RNA and proteins at the cell surface and sheds infectious virions by budding to complete the viral life cycle.

This genetically complex life cycle contains some enzymatic steps normally carried out by the host cell (such as ordinary transcription) but in addition involves extraordinary processes (such as reverse transcription) occurring rarely, if at all, in uninfected cells. Many of these unusual reactions are catalyzed by virus-encoded enzymes. Thus, a complete understanding of the viral life cycle requires a description of all the viral proteins and determination of the functional role of each protein in the various steps of replication. Direct biochemical and immunological analyses of virions have served to identify most of the virally coded proteins and the approximate structural positions of these polypeptides in both the infected cell and the mature virions. However, the function of these proteins has only been inferred from biochemical data. The analysis of viral mutants has been and continues to be of tremendous assistance in defining viral functions.

Very few of the mutations isolated to date have been mapped or even assigned to viral proteins, although characterization of existing mutants is still underway. Recombinant DNA techniques are facilitating these efforts and have made possible construction of new mutants in specific regions of the genomes of retroviral DNAs. As we now enter a new era of contact between genetics and biochemistry, our knowledge will expand enormously. This is an appropriate time to summarize our current understanding of the life cycle of these extraordinary viruses. This article will review the properties of the many mutants of murine leukemia viruses isolated to date and summarize what is known of the role of the viral gene products in specific stages in the life cycle. In addition, the nature of the interactions seen between two mutant viruses coinfecting a single cell will be discussed. Finally, brief consideration will be given to viral interactions with host genetic elements.

2 The Wild-Type Viruses: Their Genes and Genomic Organization

By far the best-studied murine leukemia viruses (MuLVs) are of the ecotropic class – those viruses isolated from mice which replicate most efficiently in mouse cells. Examples include Moloney murine leukemia virus (M-MuLV), Rauscher murine leukemia virus (R-MuLV), Kirsten MuLV, Friend-helper murine leukemia virus (F-MuLV), Gross MuLV, and the closely related AKV. All these viruses encode proteins of roughly similar size, and in many cases polyclonal antibodies directed against the proteins of one virus have been shown to cross-react with the corresponding proteins of other viruses. Nucleic acid studies also indicate the existence of extensive regions of homology between these viruses as judged from annealing of labeled DNAs to heterologous RNAs. There are also, however, at least short regions which are unique to several of the viruses: virus-specific nucleic acid probes can be generated either by subtractive hybridization protocols (STEFFEN and WEINBERG 1978; STEFFEN et al. 1980) or by the use of very short (approximately 100 bp) DNA fragments (I. Verma, personal communication; HERR and GILBERT 1982). It seems likely that the overall genomic organization and gene order will be found to be the same in all these viruses.

Mutants have been isolated from several different parents including Kirsten, Rauscher, and Moloney viruses; Moloney is gradually emerging as the most popular wild-type strain. The nucleotide sequence of the genome of Moloney, which has been determined in its entirety (SHINNICK et al. 1981), has provided the clearest definition of the genes and regulatory sequences of any of these viruses. The structural features deduced from the sequence are as follows. The 5′-proximal sequences of the integrated provirus contain the promoter for transcription by the host RNA polymerase II (pol II) system; this region shows all the hallmarks of other pol II promoters, including the TATA box, CAT box, and a "72-bp" region found to be directly repeated in many (though not all) isolates of the virus. The repeated sequence has been shown to provide "enhancer" activity to this and other promoters in experiments similar to those defining a 72-bp transcriptional enhancer of the SV40 early promoter, although no extensive sequence homology between these sequences is apparent. The entire 450-bp region of the provirus upstream of the transcription initiation site is termed the U3 region (for *u*nique sequence found at the *3*′-end of viral RNA). Immediately after the site of transcriptional initiation there is a sequence of 50 bp, termed the R region, which mediates minus-strand strong-stop DNA transfer; this is followed by a sequence of 150 bp, termed the U5 region, encoding no known function. These three sequences (U3-R-U5) together constitute the entire LTR.

Adjacent to the LTR region is the primer binding site: a sequence of 18 bp complementary to the 3′-terminal bases of proline tRNA of the mouse, which acts as a primer in initiating synthesis of minus-strand strong-stop DNA. This is followed first by the donor splice site for formation of the *env* mRNA and then by the largest region of uncertain function in the MuLV genome: 400 bp extending to the translational start site of the $Pr65^{gag}$ protein. Some of this region (possibly along with the R and U5 regions) is necessary in *cis* for viral RNA encapsidation (see p. 59).

The three major gene products are encoded in the next 8000 bp of DNA – *gag, pol,* and *env*. The *gag* gene actually encodes two major primary translation products – a glycosylated surface protein, $gPr80^{gag}$, and a soluble protein $Pr65^{gag}$, which is processed to make the four virion proteins P15, P12, P30, and P10. The information encoded by the *pol* region is apparently only translated into protein in the form of the *gag-pol* precursor $Pr200^{gag\text{-}pol}$; it is later processed to yield the mature polymerase, P75, and possibly other unidentified products. This fact introduces some complexity into the genetics of the *gag* and *pol* regions, as we shall see. The product of the third gene, the *env* protein, is synthesized from a spliced mRNA and is translated separately from *gag* and *pol*. Closely following the *env* gene is a second copy of the entire LTR sequence. The only sequences of apparent function in this region are a polypurine tract thought to be involved in plus-strand DNA synthesis, the R region utilized for minus-strand strong-stop transfer, and a polyadenylation signal targeting the 3′-end of the viral transcript to the end of the R region.

The features of M-MuLV mentioned above are likely to be universal among the MuLVs. The few cases available for which comparisons between different strains of MuLV can be made suggest that these features are conserved and therefore probably are important. For example, the very tips of the LTRs of a variety of these viruses are extremely well conserved and can be assumed to be essential for viral integration, the reaction that occurs at these sites. The locations of numerous base changes elsewhere in the genome in such viruses as Moloney and AKV can be used to define regions which are flexible and presumably less critical for replication. In all these situations, it is often fruitful to consider one of the MuLVs as a mutant or variant of another. When the biological properties of two such isolates do differ, the differences must result from one or more of the many changes in the base sequence. Appropriate exchanges of genetic material between the two viruses and analysis of the biological properties of the hybrids can be used to determine which regions are responsible for a given behavior. Examples include the amphotropic viruses. These viruses are fully replication-competent but have an expanded host range and are able to replicate in cells of many species; the difference from the ecotropic viruses responsible for this phenotype lies in the *env* region. This sort of genetic comparison of "wild-type" viruses has been one of the most productive approaches to date in assigning functions to specific DNA sequences.

The murine leukemia viruses are replication-competent but remarkably benign to cells and animals, as a rule causing essentially no cytopathic effect on cells in culture and, for a considerable period of time, causing no disease in animals. Only after a long latency period, and only in certain genetic backgrounds, do these viruses cause leukemia. To a first approximation, these viruses are harmless parasites of their mammalian hosts, capable of productively infecting an enormous number of cells without eliciting morphological transformation. Unlike the acute transforming retroviruses, therefore, the replication-competent leukemia viruses appear to carry no active oncogenes.

Despite their apparent inability to kill or transform cells, these viruses can have complex genetic effects on the host. In essence, the viruses act as mutagens: they can disrupt genes (Varmus et al. 1981; Jenkins et al. 1981) by direct insertion, they can activate genes by insertion nearby (D. Steffen, personal communication), and they can even transport genes to new cells after incorporating these sequences into the region between the LTRs (Rapp and Todaro 1978;

GOFF et al. 1980; BLAIR et al. 1981). Thus, the interactions of the viral genome with the host are particularly rich compared with most RNA viruses. The viruses also exchange genetic material among themselves at very high frequencies whenever two replicate within a single cell. Their amazing ability to exchange genetic material with genes of any source allows rapid evolution of the viruses and of any of the genes they carry.

3 Mutants: Steps in the Life Cycle Mediated by Viral Genes

Mutants of retroviruses can be generated and analyzed far more readily than mutants of most viruses for one important reason: the mutant genomes, even if absolutely defective and incapable of replication, can still be propagated as proviruses integrated into the genome of a cell. Thus, absolutely lethal mutations can be generated and the effect of the mutations can be determined without the need for helper viruses, which would otherwise be used for the propagation of the mutant. Most of the mutants discussed below were isolated in producer cell lines which serve to carry and express the mutant virus.

Some of the mutants described below were isolated after mutagenesis of the virus or of cells producing the virus. More recently, it has been demonstrated (SHIELDS et al. 1978) that M-MuLV gives rise to mutants at very high frequencies whenever the virus is transferred from one cell to another. This instability is presumably due to one of the two error-prone replication enzymes used by the retroviruses – reverse transcriptase and RNA polymerase II. Neither enzyme is thought to carry out proofreading of the product and both probably have error rates of the order of 10^{-3} (BATTULA and LOEB 1974), a frequency which would nicely account for the observed rate of appearance of mutants (GOFF et al. 1981). Once the provirus is inserted into the chromosome of the infected cell, replication is carried out by the cellular DNA polymerases, which have very low error rates, making the producer cells genetically stable. Thus, mutants can be readily generated using the unstable stage of replication but then "fixed" by maintenance of the virus in a stable stage. If viral transmission were always error-prone, reversion of mutations would be rapid, and this is not always true. Those mutations that do not readily revert may carry multiple or deletion mutations; alternatively, some base substitutions may be infrequently generated by the replicative enzymes.

3.1 Early Events: Adsorption and Uncoating of Virions

Many murine leukemia viruses have been isolated that differ in the range of species they infect. The host range of the MuLVs was one of the properties first used in their classification and is now known to reflect differences in the *env* gene and the corresponding receptor molecules on the cell surface with which they interact. The main classes are: the ecotropic viruses, best able to grow on murine cells; the xenotropic viruses, best able to grow on mink or other nonmurine (even avian) cells (LEVY 1979); the amphotropic viruses, able to grow on both murine and nonmurine cells (HARTLEY and ROWE 1976; RASHEED et al. 1976); and the polytropic (or dualtropic or MCF) viruses, also able to grow on both but differing in their antibody neutralization properties (FISCHINGER et al. 1975; HARTLEY et al. 1977). All of these viruses were isolated from wild or inbred mice, and no mutants have been derived by direct mutagenesis which show altered host range. A recent review of the properties of the xenotropic viruses is available (LEVY 1979) and they will not be discussed in detail here.

One of the most powerful blocks to infection of a permissive cell is the resistance mediated by the cell's production of the envelope glycoprotein. Whenever a cell expresses high levels of a given *env* protein (e.g., when a cell has previously been infected and is a producer of virus), infection by viruses of the same interference group is prevented. All the ecotropic viruses known, for example, are of the same interference group and presumably use the same receptor. Variants altered in the *env* gene belong to different interference groups and can bypass the block created by any particular virus.

The resistance is very efficient, allowing less than 1 in 10^6 normally infectious units of virus to enter the cell. The block probably acts very early in the infection, possibly at the time of virion binding to the receptor (BESMER and BALTIMORE 1977). The block can be bypassed if a different glycoprotein is made available to the virus for cell entry. Vesicular stomatitis viruses (VSVs) pseudotyped by MuLVs are blocked from a cell if the cell is a producer of MuLV, even if it would normally be sensitive to VSVs.

To date, four different receptors for different viruses have been genetically detected on the murine cell surface (REIN 1982; Rein, A., personal communication). The nature of the receptor molecules has not been determined in detail, although proteins which interact with the *env* protein have been studied (LANDEN and FOX 1980). It is known that mouse chromosome 5 is required for ecotropic viral infection of mouse-hamster hybrid cell lines; the gene encoding the ecotropic receptor presumably lies on that chromosome (GAZDAR et al. 1977; OIE et al. 1978; RUDDLE et al. 1978).

Other viral gene products in addition to *env* may be needed soon after infection. Several mutants have been directly isolated from wild-type parents which seem to be blocked at very early times in all normally permissive cells. The events occurring early are difficult to study biochemically because the great majority of virions applied to cells do not ultimately lead to productive infection, and analysis of the fate of virions and virion components in infected cells may not detect the biologically relevant particles. Thus, further genetic analysis of the mutants affected in these steps may prove very helpful in understanding these events.

Moloney mutant ts 1 (WONG and MCCARTER 1974) is not defective in assembly or release of particles at nonpermissive temperatures. Infection by the virions is blocked at 39° C. When cells were infected at 32° C and the temperature subsequently increased to 39° C at various times, it was found that the infection proceeded normally unless the increase occurred very soon, approximately 2 h, after infection. No further characterization of the defect has been carried out.

A startling difference in the biology between this mutant and the wild-type parent has been noted (McCarter et al. 1977). When injected into animals, the mutant does not induce the usual leukemia but instead causes a hind limb paralysis with a shorter latency period. The reason for the marked change in the pathology is unknown.

Two mutants of Rauscher MuLV, termed ts 17 and ts 19 (Stephenson et al. 1974a), were found to be very similar to ts 1. Temperature shift experiments showed that the temperature-sensitive period of infection occurred less than 8 h postinfection. Neither of these two viruses was temperature-sensitive in the *pol* function (Lai et al. 1978), and the defect probably preceded reverse transcription. Other studies of ts 17, however, showed that late steps, such as assembly, were also affected (see below, p. 57). Thus, the ts 17 mutant may be a double mutant or it may be affected in a function needed both early and late in the life cycle. The gene or genes affected in any of these mutants has not been determined.

A series of M-MuLV deletion mutants recently isolated (Crawford and Goff 1984) exhibit phenotypes similar to those of the conditional mutants. Deletions in the *gag* proteins P15 and P12 have no effect on late events and virions are released at high levels. The particles are also capable of reverse transcription in vitro. However, infection by these particles does not result in detectable DNA synthesis in vivo. Thus, some very early events such as uncoating can apparently be affected by *gag* mutations. Portions of the P15 and P12 regions are apparently dispensable in particle formation but play an important role in delivering the viral core to the appropriate part of the cell, or in an appropriate configuration, for reverse transcription to occur.

One mutation in the *env* gene has also been constructed that is blocked at an early stage. A library of insertion mutations was generated by the enzymatic insertion of 6 bp of synthetic DNA into a variety of positions in the *env* gene (Colicelli, Lobel, and Goff, manuscript in preparation). One of these mutants proved to be temperature-sensitive for the production of XC syncytia. The temperature-sensitive step occurs early in the infection; virus production is normal at both temperatures.

3.2 Reverse Transcription

Among the earliest events detectable after infection of a permissive cell is the synthesis of viral DNA. A large number of temperature-sensitive mutants have been isolated which are affected in this function. The best characterized is the Rauscher mutant ts 29 (Stephenson and Aaronson 1973). This virus can adsorb and penetrate normally but cannot productively infect cells at 39° C. The reverse transcriptase activity, both in virions and after extensive purification, was found to be thermolabile (Tronick et al. 1975). This result proved that the enzyme is encoded by the virus and that the enzyme is responsible and required for this step in the life cycle. It may be that other mutations are also present in this mutant, however. Mutant ts 29 has been shown (Van de Ven et al. 1978) to be defective in *gag* maturation as well.

It was found that a second enzymatic activity found in virions, the ability to degrade RNA in the form of an RNA:DNA hybrid duplex, was also rendered thermolabile by the ts 29 mutation (LAI et al. 1978). These results proved that the activity, termed RNAase H (for hybrid), was also virus-encoded and likely to be tightly linked to the reverse transcriptase function. These activities are in fact encoded by a single polypeptide (VERMA 1975). The ts 29 mutation was shown to be *trans*-acting: when tested for its ability to act as a helper virus for the transmission of defective sarcoma viruses this function was found to be temperature-sensitive.

Another series of temperature-sensitive mutants, starting with Moloney MuLV as the parent, was generated from an infected population of NIH/3T3 cells (GOFF et al. 1981). Clones were screened for the release of reverse transcriptase which was sensitive to incubation at 45° C; about half of these clones proved also to be temperature-sensitive in vivo. Significantly, many of these mutants chosen only on the basis of a defect in *pol* proved also to be defective in *gag* cleavage.

One spontaneous mutant of a B-tropic virus (WN1802B) was shown (GERWIN et al. 1979) to have a nonconditional defect only in the *pol* function; *gag* and *env* proteins appeared normal. Very low levels of reverse transcriptase were produced, and only aberrant *pol* precursor proteins were detected.

Numerous nonconditional mutants lacking *pol* protein and activity were found in a series of mutants isolated without mutagenesis by SHIELDS et al. (1978). Mutant M6 releases particles without detectable *pol* protein. Mutant M13 is similar, and in addition seems to show elevated *env* synthesis. This mutant has been further characterized (LEVIN and SEIDMAN 1981) as packaging reduced levels of $tRNA^{pro}$ into virions; the reduction, however, was in the free, unbound tRNA pool of the virion and the tRNA bound to genomic RNA was present at normal levels. The *pol* protein may, then, play some role in the accumulation of tRNAs in virions but is not absolutely required. Mutant M23 contains a 1.4 kb deletion largely, if not entirely, within the *env* region but may have additional defects in *pol*. This producer cell line releases particles at somewhat reduced levels, and no *pol* or *env* proteins are found in the virions. This line has been useful as a special host: since only *gag* is produced, and since the line is superinfectable by all ecotropic viruses, it can support the growth of other mutants affected only in *gag* (GOFF et al. 1981; SCHWARTZBERG et al. 1984; CRAWFORD and GOFF 1984). Three other mutants (NX-2, NX-3, NX-4) isolated after infection of normal rat kidney (NRK) cells were characterized (YOSHIMURA and YAMAMURA 1981) as defective in the *pol* function. In these cases also, particles were released but no reverse transcriptase could be detected.

A large array of mutations in the *pol* gene was generated recently by the in vitro mutagenesis of a cloned DNA copy of the M-MuLV genome (Schwartzberg et al., manuscript in preparation). All of the mutants with alterations in the 5′-portion of the *pol* region showed a similar phenotype: uninfectious virions were released, completely lacking in reverse transcriptase activity. Like many of the temperature-sensitive mutants described above, these mutants also showed a reduced rate of cleavage of the *gag* precursor. These mutants support the notion that parts of the *pol* region are required for *gag* maturation and constitute stronger evidence than results with the temperature-sensitive mutants because double mutations cannot be present. Mutations mapping in the 3′-portion of *pol* did not affect reverse transcriptase. A second class of mutations which affected reverse transcriptase function was also found. Deletions in the P30 domain of the *gag* gene, even though they were not frameshift mutations and thus did synthesize the *gag-pol* precursor, were unable to provide *pol* function. When introduced into M23 cells, which constitutively produce *gag* protein and release particles, these P30 mutants could not contribute *pol* to the particles. Thus, regions of the *gag* gene must be expressed in the *gag-pol* precursor for reverse transcriptase to be incorporated into particles and matured (SCHWARTZBERG et al. 1984).

Presumably, *cis*-acting mutations blocked at various stages of reverse transcription could also be found. These mutations would be of great interest as defining regions needed for initiation of synthesis of each DNA strand and for strong-stop transfers or "jumps." Sequences upstream of the start of either strong-stop DNA strand, and in the R region, are likely targets for mutagenesis. No mutations of this type have yet been reported.

3.3 Integration of the Viral DNA: Establishment of the Provirus

Because of the importance of integration of the viral DNA, both to the virus and the host cell, there is great interest in understanding the details of this recombination reaction. No biochemical data have been obtained which could shed light on this process. The structures formed at the sites of recombination, however, have been studied in some detail. The provirus insertion generates a 4-bp duplication of the sequence at the target site; these base pairs flank the inserted viral DNA. Such duplications are hallmarks of insertions mediated by bacterial and eukaryotic insertion sequences and transposable elements. Aberrant reactions have also been observed and characterized (SHOEMAKER et al. 1980, 1981a, b) in which intramolecular integrative recombination reactions occur on the preintegrative viral DNAs. These reactions generate inversions and deletions of viral sequences. If the products of such reactions can subsequently integrate into the host, then these reactions may be an important cause of viral mutations. If such reactions can occur to normal proviruses after insertion into the host, they may be a major cause of mutations in the neighboring host DNA. These possibilities have not yet been explored.

Several viral mutants unable to form the integrated provirus have recently been constructed (Schwartzberg et al., manuscript in preparation). Of a series of deletion mutations generated at various positions in the *pol* gene of M-MuLV, a number were found to leave virion formation and reverse transcriptase activity unaffected. These mutations were clustered at the 3′-end of the gene. Viral DNA could be inserted into cells by DNA-mediated transformation, and these cells became producers of virions complete with reverse transcriptase activity. When cells were infected with these mutant virions, viral DNA was formed normally and all three species (linear and two supercoiled) of DNA were detected, but the cells did not become producers of virus. Apparently, the DNA could not efficiently generate the productive structures readily formed by DNA-mediated transformation. Similar point mutations in the same region have been generated in another laboratory (Varmus et al., personal communication). The mutations can revert to the wild type after long-term passage by recombination with endogenous viral sequences (Schwartzberg et al., manuscript in preparation).

The above mutations must affect an unidentified viral protein, perhaps one cleaved from the *gag-pol* precursor, which is involved in the insertion of the provirus and thus should be *trans*-acting. Other mutants have been formed with alterations in the DNA sequences presumably recognized for this reaction: the inverted repeats at the tips of the LTRs. A single base-pair change has been made by oligonucleotide site-specific mutagenesis (Colicelli and Goff, manuscript in preparation), and the resulting virus shows a phenotype very similar to the 3′-*pol* mutants. As before, DNA is made but the cells only poorly become producers of virus.

An extremely interesting and complex mutant, clone 8A, was isolated by Rein et al. (1978). This mutant of M-MuLV is able to assemble virions, and the virions can carry out at least some reverse transcription. They are XC-negative, however, and require weeks to turn successfully an infected culture into producers. Thus, these mutants may be affected quantitatively in reverse transcription, integration, or some other early event. The phenotype of this mutant is very similar to that of the 3′-*pol* deletions described above.

Host functions are probably important in the integration reaction. Numerous genes of the host have been identified which act to restrict the ability of various strains of MuLVs to induce tumor formation; the most well-known are the *Fv* genes which restrict the course of *F*riend *v*irus disease. The mode of action of many of these genes is not understood, and analysis is difficult because many of the genes do not restrict the replication of the virus in culture cells. One gene, however, which does act in culture is the Fv-1 gene of the mouse. This gene can block the replication not only of Friend virus but also of a wide variety of other related retroviruses (Lilly and Pincus 1973). Two alleles (N and B) are known: Fv-$1^{N/N}$ mice replicate N-tropic viruses, Fv-$1^{B/B}$ mice replicate B-tropic viruses, and Fv-$1^{N/B}$ mice replicate neither. The resistance is therefore dominant over sensitivity. Phenotypic mixtures (see below, p 62) of N- and B-tropic viruses are restricted by *both* alleles (Rein et al. 1976). Finally, there exist NB-tropic viruses, such as Moloney MuLV, which are restricted by neither allele. In all cases, the restriction is not absolute, but leaky and probably multiplicity-dependent. The restriction occurs at a remarkable time in the course of infection: after the synthesis of the viral DNA but before the integration of the DNA into the host genome (Jolicoeur and Baltimore 1976). This conclusion is supported by earlier genetic data (Huang et al. 1973) showing that Fv-1 cannot block infection by VSV contained in MuLV coats; Fv-1 does not act at the cell surface.

A complete description of the Fv-1 system is beyond the scope of this review, and in fact a detailed report filled a chapter in a previous volume in this series (Jolicoeur 1979). Recent results have been reported, however, which allow exact mapping of the viral gene important in the restriction to replication. DNA fragments were exchanged between two cloned viral genomic DNAs, one N-tropic and one B-tropic, and the tropism of the recombinants was determined (P. Jolicoeur, personal communication; BOONE et al. 1983). The important region of the genome was the P30 region of the *gag* gene, in agreement with earlier mapping data (HOPKINS and JOLICOEUR 1975; HOPKINS et al. 1977a, b; FALLER and HOPKINS 1977, 1978; SCHINDLER et al. 1977). Sequence analysis showed that viruses able to replicate on N, B, or NB mice differed at only one or two amino acids in the P30 region. These results prove that P30 carries out important functions after entry into the cell and may even implicate the protein in the integration of the proviral DNA into the host chromosome in a direct way. How this reaction is carried out and the relationship of the P30 protein to the 3′-pol protein are not at all clear.

3.4 Virion Assembly and Release

By far the most commonly isolated mutants have been those defective in the formation of virion particles and release into the extracellular fluid. This may reflect the ease of detection of such a phenotype, or it may be due to the possibility that many viral proteins are required for the process and that the target size for the mutagenic event is large. Some of the first temperature-sensitive mutants of mammalian retroviruses were mutants of Kirsten MuLV (STEPHENSON et al. 1972), Moloney MuLV (WONG and MCCARTER 1973, 1974; WONG et al. 1973), and Rauscher MuLV (STEPHENSON and AARONSON 1973, 1974).

One set of mutants was made by bromodeoxyuridine or N-methyl N-nitro N-nitrosoguanidine mutagenesis (STEPHENSON et al. 1972). Of many temperature-sensitive mutants, the Kirsten MuLV mutants ts6 and ts9 (STEPHENSON and AARONSON 1973) were found to be incapable of the production of infectious virus when a producer cell was shifted to the nonpermissive temperature. Mutants ts3 and ts7 were also late mutants; no particles were released at the high temperature. All reverted at very low frequencies. No further characterization has been reported.

The Moloney mutants were isolated by selecting for an inability of the virus to induce flat cells, termed S^+L^-cells, to become morphologically transformed under the nonpermissive conditions (WONG et al. 1973). These cells harbor an inactive sarcoma provirus which is insufficiently expressed (for unknown reasons) to elicit morphological transformation. Upon superinfection by replication-competent retroviruses, the resident sarcoma provirus is rescued and can then induce transformation, presumably by reinfection of cells in the population. Mutations in any function needed for rescue will render the retrovirus negative in this assay. Six mutants were selected; reversion frequencies were determined to be in the range of 1 in 10^4 (WONG and MCCARTER 1974). Upon shifting the producer cell lines carrying one of these mutants, M-MuLV ts3, to the nonpermissive temperature, particle release could be shown to stop (WONG and MCCARTER 1974), and the intracellular precursors $Pr65^{gag}$ and $Pr200^{gag\text{-}pol}$ accumulated without the normal proteolytic cleavage (WITTE and BALTIMORE 1978). Bud structures were observed at the nonpermissive temperature (WONG and MCCARTER 1974; WONG and MACCLEOD 1975; LU et al. 1979); immature virions could be collected immediately after a temperature decrease, and processing of the *gag* precursor could be correlated with changes in the morphology of the particles in the electron microscope. This mutant has been exceedingly useful in structural analyses of the budding process (YUEN and WONG 1977). The maturation was shown to occur even in the presence of cycloheximide (WITTE and BALTIMORE 1978) and was very rapid after the temperature decrease. It should be noted that the wild-type virus may not show the same behavior as the mutant in the kinetics of $Pr200^{gag\text{-}pol}$ cleavage (YOSHINAKA et al. 1980). Interestingly, at the high temperature, there was also found to be a rapid turnover of the *env* precursor in this mutant (WITTE and BALTIMORE 1978).

A third early isolation of mutants was carried out using Rauscher MuLV as the parent virus. This screen (STEPHENSON and AARONSON 1973) depended only on differences in XC cell syncytium formation by duplicate cultures assayed at different temperatures. Mutants ts17, ts24, ts25, and ts26 showed phenotypes very close to that of M-MuLV ts3 and less extensive data suggested that mutants ts20, ts21, ts23, and ts27 might be very similar. In these cases too, proteolytic processing of $Pr65^{gag}$ and $Pr200^{gag\text{-}pol}$ were reduced, and particle formation was blocked at the nonpermissive temperature (STEPHENSON et al. 1975; VAN DE VEN et al. 1978; WITTE and BALTIMORE 1978). The formation of buds after temperature elevation and under different conditions of sample preparation for mutant ts25 has been studied in some detail (DEMSEY et al. 1979). It was also shown that if particles of mutants ts25 and ts26 were allowed to form at the permissive temperature, then these particles could carry out early events normally at the nonpermissive temperature (STEPHENSON and AARONSON 1974). Some of these mutants may be blocked at slightly different stages; bud structures were not seen with some of the mutants (YEGER et al. 1976, 1978) at the high temperature as they were in ts24 and in Moloney mutant ts3 (LU et al. 1979). However, the extreme dependence of the budding structures seen in the electron microscope on the conditions used for the sample preparation should be kept in mind (TRAKTMAN and BALTIMORE 1982; YEGER et al. 1978). These mutants were first exploited to deduce the order of the mature cleavage products in the *gag* precursor: partial cleavage products were analyzed to determine the linkage of each product with its neighbors (REYNOLDS and STEPHENSON 1977). In agreement with later physical data, the order was found to be P15-P12-P30-P10.

Mutant ts26 was later shown (RUTA et al. 1979) to have an additional alteration in *env* processing. At the high temperature, *env* protein did not mature, and the cell surface became depleted of protein; virions were subsequently released without the glycoprotein. This may be the primary lesion in this mutant. These results suggest that *env* protein is not needed for bud formation and release of virions, but may imply that *env* has some effect on the rate of *gag* cleavage.

Another mutant of the R-MuLV series, ts28, was normal in cleavage of $Pr65^{gag}$ and release from the cell; however, the particles were not infectious

(Van de Ven et al. 1978; Stephenson and Aaronson 1973) and exhibited distorted particle morphology (Yuen and Wong 1977). Similarly, a mutant, ts7, has been described (Wong and Gallick 1978) which produces uninfectious particles at the nonpermissive temperature; the particles themselves are somewhat thermolabile.

A series of nonconditional variants of R-MuLV have also been described (Sacks et al. 1978) which are unable to process $Pr65^{gag}$.

Besmer et al. (1979) have characterized a spontaneous mutant of Moloney MuLV which is unable to express normal *gag* or *pol* proteins, but does make normal *env* precursor. No particles were released from this variant. The defective genome was shown to contain a 1-kb deletion, which does not affect *env* mRNA size or function.

Collins and Chesebro (1981) have also characterized a mutant defective in virion assembly in which the *gag* precursor was poorly cleaved; aberrant *pol* peptides were seen, and *env* incorporation into virions was limited to low levels. The parent virus was the Friend-helper MuLV.

The gene or genes actually mutated in any of these mutants is not known, although the *gag* region is plausibly a common target. The mutants could be directly affected in precursor-precursor interactions (i.e., in capsomer interactions), or in the ability of the *gag* precursor to fold, be transported, or be cleaved properly to the four mature polypeptides. This explanation for the observed phenotype is supported by complementation experiments. When a distinct xenotropic virus was introduced into the same cell harboring the mutants, no production of the ecotropic viral proteins was observed at the nonpermissive temperature (Reynolds and Stephenson 1977). Thus, the defect in these cases is likely to be in the ability of the viral protein to assemble or be cleaved, rather than in a *trans*-acting enzyme needed for catalysis of assembly. The close coincidence of the two phenotypes – failure to assemble and bud on the one hand, and failure to cleave the precursors on the other – lends credence to the notion that budding and cleavage are obligatorily coupled. One mutant (Collins and Chesebro 1981) may be an exception to this rule. An interesting observation was recently reported (Traktman and Baltimore 1982) that treatment of cells at the nonpermissive temperature with a variety of proteases can bypass the mutational block and allow release of infectious particles.

New mutants made by direct alteration of the *gag* gene show the importance of the gene for assembly and, furthermore, pinpoint the crucial domains. Deletions of P15 and P12 show very little effect on assembly and release (Crawford and Goff 1984) and are presumably not intimately involved in capsid interactions. However, deletions of P10 have some quantitative effects on assembly and release of particles, and deletions of P30 totally abolish these processes. Presumably these domains play important roles.

Two other mutants have been isolated which are conditionally affected in cleavage and release, along with other phenotypes. R-MuLV mutant ts17 (Stephenson and Aaronson 1973; Yeger et al. 1976; Van de Ven et al. 1978) accumulates uncleaved $Pr65^{gag}$ at the high temperature; however, the mutation has a second effect and blocks the ability even of virus harvested at the low temperature to initiate an infection at the high temperature. The second mutant, ts29, is slow to cleave *gag* at the nonpermissive temperature. Aberrant bud structures have been noted in cells infected with this mutant (YEGER et al. 1976); *env* protein is associated with these arrested structures (YEGER and KALNINS 1978). It is curious that the primary lesion detected in this mutant is in *pol:* The reverse transcriptase is temperature-sensitive (see above, p. 51). Similar M-MuLV mutants appeared in a screen which isolated mutants with a temperature-sensitive *pol* function (GOFF et al. 1981). Two out of four such mutants were also found to be defective in cleavage of *gag*.

Several nonconditional mutants affected in assembly have been isolated by SHIELDS et al. (1978). These viruses, described above as defective in *pol* function, are capable of causing release of particles but show a reduced rate of cleavage of $Pr65^{gag}$; examples include M6, M13, NX-2, NX-3, and NX-4 (YOSHIMURA and YAMAMURA 1981).

Some genetic analysis has placed the ts25 lesion near the 3′-end of *gag* or the 5′-end of *pol*. This result is made especially interesting by the recent finding that there is space for the coding of a protein immediately 3′ to the *gag* gene but 5′ to the region encoding reverse transcriptase (S. Oroszlan, personal communication). This position in the avian retroviruses has been shown to encode a protease required for the cleavages in the *gag* precursor. Thus, the ts25 lesion could possibly be affected in a mammalian equivalent of the avian protease, and this protease could map near the 5′-end of *pol*.

Recently, a number of deletion mutants were constructed in this laboratory (Schwartzberg et al., manuscript in preparation) at various positions within the *pol* gene. All of these mutations caused a drastic reduction in the rate of cleavage of $Pr65^{gag}$. Thus, it seems very likely that some part of the *pol* gene will prove necessary for *gag* processing, either to help fold the *gag* protein into the correct conformation or directly to make the proteolytic cleavages. The *gag* and *pol* products are interdependent: portions of the *gag* gene are absolutely necessary for *pol* expression and utilization; portions of the *pol* gene are necessary for proper processing of the *gag* protein. This interrelationship may well account for the complicated pleiotropic behavior of so many of the temperature-sensitive mutants. In the case of the mutants derived in vivo, however, it is entirely possible that multiple mutations exist and are responsible for the multitude of effects.

3.5 Mutants Blocked in Packaging Viral RNA into Particles

Recently, a deletion mutant of M-MuLV was constructed (MANN et al. 1983) which produced normal levels of virion particles and normal levels of reverse transcriptase. The particles proved to be unable to transmit the viral genome to infected cells, and examination of the nucleic acid content of the virions revealed that genomic RNA was completely absent from the released particles. The sequence deleted was thus apparently necessary for recognition of the RNA by some virion protein responsible for the selective packaging of viral genomic RNAs. No cellular RNAs replaced the viral RNA as was found in the case of an analogous avian mutant (LINIAL et al. 1978).

The existence of this mutant suggests that RNA plays no part in any structural way in virion protein assembly and in the incorporation of reverse transcriptase by particles. The proteins formed by the mutant were shown to be fully functional. The mutation was *cis*-acting and not *trans*-acting: other viral genomes can be successfully packaged into the mutant coats. Thus, the mutant is useful as a means of transferring any defective virus (such as a transforming virus or a viral vector which carries a cellular gene of interest) without also transferring the helper virus genome; i.e., this virus is the perfect helper for the generation of nonproducer cell lines. It will certainly also be of use in the transfer of genes into various tissues of the mouse and into the germ line, when it is important that the animal not be exposed to wild-type virus.

The sequences deleted in this mutant lie between the left LTR and the start of the *gag* gene; the exact extent of the sequences essential for packaging, however, is not yet known. A large region of M-MuLV, about 200 bases from the Pvu I site to the start of the *gag* gene, is in fact known to be unnecessary in packaging (SCHWARTZBERG et al. 1983); it is likely that the important region is very close to the LTR. Moreover, the actual viral protein which recognizes this sequence and collects the RNA into the particle is not yet known. No mutants in the *gag* or *pol* region have been isolated which are defective in the incorporation of wild-type viral RNA into particles. One or more of the *gag* proteins, however, are likely to be responsible. A cell line termed M23 (SHIELDS et al. 1978), expressing only the *gag* polyprotein and neither *pol* nor *env* products, has been shown to release particles (albeit at reduced levels) and selectively to package viral RNA into them.

3.6 Leukemogenesis and Tissue Specificity

The mechanism or mechanisms of leukemogenesis by the MuLVs is not known with certainty, but recent results have reduced some of the mystery. It is clear that at least some tumors are initiated by the activation of the *c-myc* oncogene through the insertion of an MuLV provirus nearby (D. Steffen, personal communication). If this proves to be the only or most common mechanism, we need to understand the circumstances under which this oncogene can be activated and what subsequent events might be required to lead to the transformed phenotypes. The role of the recombinant viruses known as mink cell focus inducing (MCF) viruses in this process, described in a recent review (FAMULARI 1983), is still not known. The analysis of viral and mouse variants with different interactions is almost the only productive approach to this problem to date.

One set of variants, perhaps derived from a single parental virus, has been isolated which includes viruses that induce leukemia with different efficiencies (MANLY and BUFFETT 1979). Several distinct isolates of replication-competent retroviruses show extensive homology to the leukemogenic viruses discussed above but fail themselves to induce disease in animals (JOLICOEUR et al. 1978; reviewed by GROSS 1970). These viruses thus contain some genetic differences which do not affect their growth in fibroblasts in culture but reduce their ability to transform lymphocytes. There is evidence that the leukemogenicity of these viruses correlates closely with their thymotropism (HAAS 1978; CLOYD et al. 1980). Recently, DNA sequences have been exchanged between molecular clones of two viruses, one thymotropic and one not, in an attempt to localize the sequences responsible for the phenotype of thymotropism (DESGROSEILLERS et al. 1983). The U3 region of the viral LTR was found to carry the determinants of the tissue specificity. This result suggests that transcription of the viral DNA may be the tissue-specific process. The major difference between the two U3 regions is the existence of the 72-bp "enhancer" region: The thymotropic B-Cl-11 virus carries a tandem duplication of its enhancer (as does M-MuLV); the thymotropic Gross passage A virus carries an insertion of novel DNA in its unique enhancer region; and the nonthymotropic N-Cl-35 virus shows only one copy of an unmodified enhancer. Thus, there may exist thymus-specific enhancers which are needed to allow efficient replication of the virus in the organ, and it may be that this replication is an essential prerequisite for leukemogenesis.

Similar results have also been obtained in the localization of the determinants of tissue tropism that differ between M-MuLV and F-MuLV (CHATIS et al. 1983). In this case too, the thymotropism of M-MuLV was found to lie in a small DNA fragment containing the 3′-end of the *env* gene and the U3 region of the LTR. Thus, the targeting of M-MuLV to T cells and of F-MuLV to erythroid precursors may be due to the tissue-specific transcription of the provirus rather than to a specific alteration in a viral protein. It cannot be ruled out yet that the *env* gene may play some role.

These results bring together the classic C-type viruses described here with the structurally distinct mouse mammary tumor viruses (MMTVs). These viruses are now seen to be similar in showing tissue-specific transcription. The MMTVs have evolved an even more specific transcriptional promoter: RNA is synthesized only in cells with a receptor for glucocorticoid hormones and only when the cells are exposed to the hormone. The result is an extreme tissue tropism. The two classes are similar in another way: the mechanism by which the leukemia is induced. It has recently been shown that both these viruses can act in some tissues to activate an endogenous oncogene (NUSSE and VARMUS 1982; D. Steffen, personal communication). Thus, the great difference in the nature of the diseases caused by these viruses may be largely a consequence of the different transcriptional promoters which they utilize.

3.7 Silent Mutations

Several mutations have been generated which have a minimal effect on the replication of the virus. Although seemingly uninformative, these mutants serve an important role in identifying flexible portions of the genome. When the mutation affects a protein, it can be deduced that the region affected is dispensable; when a noncoding region is altered, the sequences are apparently not recognized by any protein and can serve to delineate the boundaries of a binding site.

The largest region of the genome of M-MuLV found to be dispensable is the sequence near the 5′-end of the *gag* gene. Large deletions in this region

(SCHWARTZBERG et al. 1983) permit the viruses to grow as well as wild-type viruses in fibroblasts and lymphocytes and also cause leukemia in animals with normal kinetics (N. Rosenberg, personal communication). These viruses lack the glycosylated *gag* protein, or in some cases synthesize a nonglycosylated variant; this protein is, therefore, unnecessary for any known viral function. Similar results have been reported (FAN et al. 1983) with a point mutation in this region. A collection of oligonucleotide linker insertion mutants yielded many viable viruses (Lobel and Goff, in press); each of these defines a mutable region. The commonest sites for these insertions are within the viral LTR, implying that much of the sequence is not needed for such functions as promotion of transcription or that the promoter is highly flexible. Interestingly, a short probe has been isolated from the LTR (I. Verma, personal communication) that is very specific for Moloney MuLV and that shows poor hybridization to other viruses. Thus, this region may be able to diverge and evolve comparatively rapidly.

Finally, numerous mutations have been isolated which do not affect the ability of the virus to replicate in permissive cells but which abolish or reduce the ability of the virus to induce XC syncytia (HOPKINS and JOLICOEUR 1975; RAPP and NOWINSKI 1976; NOWINSKI et al. 1977; REIN et al. 1979a). The mutants include isolates derived from BALB/c mice, C3H/10T1/2 cells, and several M-MuLVs (mutants M10 and M11) isolated after viral transmission (SHIELDS et al. 1978). None of the mutations have been mapped, and the gene affected is not known. Apparently, the ability to induce syncytia is not a requirement for replication and is easily separable from it. Isolates of amphotropic and MCF viruses are XC-negative (NOWINSKI et al. 1977); since these viruses differ from ecotropic viruses in the *env* gene, it may be that the *env* protein is an important part of the induction of XC cell fusion.

4 Genetic Events Mediated by Retroviruses

Because the retroviruses direct the synthesis of DNA and covalently insert that DNA into the host cell, a rich repertoire of genetic events is mediated by these viruses. Two viruses can undergo recombination at various stages of the life cycle; genes can be acquired from the cell by a virus and transduced to newly infected cells; genes can be interrupted or activated by insertion of the virus into the genome.

4.1 Phenotypic Mixing and Complementation

Two retroviral RNA genomes produced within a single cell can generate viruses which contain proteins and RNA of both of these genomes. The commonest and most extreme example of the process is the rescue of a defective virus by a fully replication-competent wild-type helper; in these cases, all of the virion proteins are usually provided by the helper. The defective genome is packaged along with the helper genome to yield a mixed virus population. Obviously, this process is exceedingly useful in allowing the manipulation and transfer of defective viruses between cells. Until recently, such procedures were the only means of passing defective genomes to new cells; the existence of *cis*-acting packaging mutants (Mann et al. 1983; see above) now allows the preparation of virus stocks containing only the defective genome.

The genomic RNA in retroviral particles is in a dimeric structure. When two viral genomes are packaged by the same cell, they might in principle be restricted to the formation of homozygotes or they might be able to form heterozygotes within a single virion; no definitive biochemical data have resolved these possibilities. Good genetic evidence suggests that heterozygotes do form (McCarter 1977). It is likely, moreover, on the basis of the very high level of recombination that has been observed between two such genomes (Stephenson et al. 1974a, b; Goldfarb and Weinberg 1981) that heterodimers can form.

It has also been shown that two replication-competent viruses can contribute proteins to virions and together assemble phenotypically mixed particles. In one very elegant experiment, phenotypic mixing was used to probe the dominance relationship of virions of mixed N- and B-tropism (Rein et al. 1976). A mixed virus stock was prepared and tested for growth in Fv-$1^{N/N}$ or Fv-$1^{B/B}$ cells. Unlike either parent virus, the growth of the mixed virus was restricted in both cell types; the mixture of the two P30 types presumably rendered the virus subject to attack by either allelic gene product.

The range of viruses that can interact in this way is normally small: avian viruses such as RSV(-)alpha do not cooperate with the mammalian viruses (Weiss and Wong 1977). But most of the mammalian viruses can interact, and when only the envelope proteins are considered, the range is surprisingly large. Xenotropic and ecotropic viruses readily form pseudotypes (Besmer and Baltimore 1977; Ishimoto et al. 1977), and viruses as different as vesicular stomatitis virus (VSV) can be provided with a new envelope protein (Zavada 1972; Huang et al. 1973) to allow infection of cells (Witte and Baltimore 1977). Only some *env* protein of the MuLV is provided; much of the VSV assembly machinery must be intact.

When both retrovirus genomes are defective, the two can complement to create nondefective particles. Few examples in the mammalian viruses have been reported, in part because the high rate of recombination to yield wild-type virus makes clean analyses of complementation difficult. One important example is the complementation plaque assay derived by Rein and Bassin (1978). In this system, XC-negative amphotropic viruses and a defective mutant of XC-positive Moloney MuLV can give rise to XC plaques when introduced into cells by coinfection. This system allows quantitation of particles of a defective virus. The XC-negative amphotropic virus can also be used to rescue such defective mutants (Rein et al. 1979b) and pass them to newly infected cells. Another similar approach has been exploited beginning with a cell line already containing and expressing a defective provirus and subsequently introducing the second virus by infection. The M23 cell line described above is ideal for this purpose, expressing only the *gag* gene (Shields et al. 1978). In this way, viruses have been isolated (Goff et al. 1981) that induce XC syncytia only on the host providing the complementing function and not on uninfected NIH/3T3 cells. Similarly, mutants have been constructed by altering cloned DNA copies of M-MuLV in vitro, and the DNAs could then be tested for complementation in the M23 cell line. Mutants which provided the *pol* and *env* functions were complemented by the *gag* function of M23 and could grow to yield XC plaques. Mutants which did not provide *pol* and *env* functions could not be complemented; importantly, when introduced as DNA they also did not recombine with the endogenous genome in M23 to generate wild-type virus (SCHWARTZBERG et al. 1984). This result implies that homologous recombination does not readily occur between endogenous DNA and exogenously added DNA, a fact well established in other systems.

4.2 Genetic Recombination Between Viruses

The generation of wild-type recombinants by coinfection of cells with two mutants has been studied for many years. The temperature-sensitive mutants described above were used in very early experiments to demonstrate the high frequency (in the range of 10^{-1}) of recombination between markers (WONG and MCCARTER 1973; STEPHENSON et al. 1974a, b). Passage of mixed cultures at high temperatures resulted in the rapid appearance of temperature-resistant virus. Mutants of two different parent viruses which carry silent polymorphisms in viral gene products have also been recombined (AARONSON and BARBACID 1980; Traktman and Baltimore, manuscript in preparation). Analysis of the wild-type recombinants has demonstrated the linkage of various genes with the selected wild-type allele and has allowed limited genetic mapping of the original lesion. Similar studies have been utilized to localize the region conferring other phenotypes on the genome, i.e., to establish linkage with a gene as measured by a biochemical marker. The Fv-1-specific tropism of viruses was first linked in this way to the P30 region of the *gag* gene (HOPKINS and JOLICOEUR 1975; HOPKINS et al. 1976, 1977a, b; FALLER and HOPKINS 1977, 1978; SCHINDLER et al. 1977); a linkage proved later to be correct. The GIX antigen was shown to cosegregate with the *env* region (O'DONNELL and STOCKERT 1976).

Recombinants also arise between exogenous and endogenous genomes whenever replication of the viruses can occur. Such recombinants can be detected by the acquisition of new host ranges provided by xenotropic *env* sequences which are endogenous in mouse cells (STEPHENSON et al. 1974b). The best-documented example is the generation of mink cell focus inducing-(MCF) viruses during replication of leukemia viruses in whole animals. A full discussion of these recombinants is beyond the scope of this review but can be found in a recent article (FAMULARI 1983). In brief, these viruses consist of xenotropic envelope sequences (derived from the host genome) joined to the parental ecotropic virus, such that a virus with a new host range results. Many such recombinants have been mapped in detail (LUNG et al. 1980). The time and mechanism of the recombination is not yet known; indeed, the role played by these viruses in leukemogenesis is obscure. If oncogene activation by promoter insertion is the major leukemogenic event, then the role of a new envelope protein might be to provide the virus access to new target cells. It may also be that the new envelope is incidental to the leukemic event but that another closely linked sequence, such as the U3 region of the LTR, plays a larger role (KHAN and MARTIN 1983).

In some cases, the recombination of mutant viruses with endogenous sequences is a serious complication to the experimental manipulation of a mutant. For example, the packaging mutant described above has been shown (MANN et al. 1983) to revert to the wild type, allowing the spreading of virus throughout a culture intended to be virus-free. Similarly, extensive passage of cells infected with a deletion mutant blocked in provirus establishment (Schwartzberg et al., manuscript in preparation) results in repair of the deletion and generation of wild-type virus; in this case, the DNA which restored the function was shown to be distinguishable from the parent virus sequence and must have derived from the cell's collection of endogenous virus genomes.

A detailed characterization of recombinants derived during a particular cross has been carried out (AARONSON and BARBACID 1980). R-MuLV mutant ts25 was copassaged in tissue culture cells with the xenotropic BALB/c 2 virus; recombinants were selected to carry the ecotropic *env* gene of the R-MuLV parent and the temperature resistance of the xenotropic parent. Analysis of independent recombinants showed that multiple crossovers had often occurred. The only region consistently donated by the xenotropic parent included the 3′-end of the *gag* gene (near P10) and the 5′-end of the *pol* gene, implying that the original lesion in ts25 might lie in this area. Further linkage studies showed that the Fv-1 tropism of the progeny cosegregated with P30, in agreement with other results.

The actual mechanism and time in the life cycle by which recombinations occur in retroviruses is not known. Data on the avian viruses (WYKE et al. 1975) suggest that heterozygote formation is important to the high-frequency recombination observed, and that "copy choice" template switches by reverse transcriptase during DNA synthesis may be very common (JUNGHAUS et al. 1982). No experiments in the murine system have directly addressed the question of whether these events are truly necessary.

4.3 Recombination Between Virus and Host Genes

An extremely important evolutionary process carried out by retroviruses is the acquisition and transduction of cellular genes. This process is the means by which acutely oncogenic transforming viruses arise from the leukemia viruses, and related processes may be responsible for cellular "pseudogenes" (LUEDERS et al. 1982). The remarkable feature of genes found on acutely transforming viruses is that they lack the introns normally present in the parental cellular gene; that is, they are cDNA copies of cellular genes. An implication of this fact is that very large cellular genes can be accommodated on a viral genome of limited size. Abelson MuLV, for example, represents the transduction of a gene spanning approximately 40 kb of cell DNA (GOFF et al. 1980; WANG et al. 1984) on a virus of less than 6 kb.

The generation of these recombinants is an extremely rare event, and no one-step culture system has been devised which reproducibly gives rise to such viruses. Structural analysis of the final recombinants, however, has led to a model for their formation (VARMUS 1982) which is generally accepted. The initial step is presumed to be the integration of a provirus upstream from the gene to be acquired. Subsequently or concomitantly, a deletion must occur which removes the 3′-portion of the provirus and fuses the remaining 5′-portion to the 3′-part of the cellular gene. Analysis of the 5′-proximal junction of transforming viruses, such as the Abelson MuLV (A-MuLV), lends some support for this event: 4 bp of homology are found between the M-MuLV genome and the c-*abl* gene at the points of recombination (WANG et al. 1984). The fusion would result in the production of a fusion transcript initiated in the viral 5′-LTR and proceeding into the cellular gene, and the 3′-end of the RNA would be determined by the cellular sequences 3′ to the gene. The hybrid RNA would be spliced at this stage to remove the gene's intron sequences. Such an RNA would serve as the substrate for formation of the final virus. Elegant experiments (GOLDFARB and WEINBERG 1981) have shown that such an RNA is efficiently packaged into virions and, furthermore, can recombine with a helper genome to append a 3′-viral LTR and form a fully transmissible virus. This latter recombination can occur in a single round of infection and seems not to require the benefit of any homology. The frequency of this recombination is quite high – of the order of 10^{-4} – 10^{-5} per infecting virion (GOFF et al. 1982).

Experimental systems to analyze these events further may be possible. A-MuLV was generated in the laboratory in 1969 by injecting M-MuLV into mice; among 159 mice, one carried out the recombination, but any intermediates in the process were not recovered (ABELSON and RABSTEIN 1970). Rapp and co-workers have been able to generate such recombinants wholly in tissue culture by long-term passage of helper virus and selection for transforming activity (RAPP and TODARO 1978). This procedure could allow identification, in principle, of all cellular genes capable of activation to induce morphological transformation. Most helpful in the analysis of the mechanism of the process would be the isolation of the postulated intermediate in which gene fusion and activation have occurred, but the final virus formation has not.

4.4 Mutation of Cellular Genes

Because the retrovirus DNA is inserted with high efficiency into the cellular genome, it might be supposed that the virus could act as transposons or phage mu do in bacteria and create mutations in cell genes. This event has in fact been amply documented in many laboratories. The most direct demonstration was the selection from a population of cells transformed by Rous Sarcoma Virus (RSV) of flat revertants after infection of the population with M-MuLV (VARMUS et al. 1981). Examination of about 60 such revertants demonstrated that in two lines the M-MuLV genome had integrated into the RSV provirus so as to inactivate the expression of the *src* transforming gene. The event was not frequent – it was in fact well below the spontaneous rate of mutation – but it was detectable, at least when probes for the target gene were available.

Recently, analysis of a cell line transformed by A-MuLV revealed the existence of a mutation caused by insertion: the cellular gene encoding p53, a transformation-related protein, was found to be interrupted by the insertion of an M-MuLV provirus (V. Rotter, personal communication). The mutation had apparently been gratuitously induced during the transformation of the cell by the helper M-MuLV used to introduce the A-MuLV genome. The effects of this mutation on the biology of the cell line are under investigation.

In another example, recent work in our laboratory and that of M.C. Nguyen-Huu has demonstrated the integration of M-MuLV into the HGPRTase gene among many 8-azaguanine- and 6-thioguanine-resistant colonies isolated after infection of F9 teratocarcinoma cells (manuscript in preparation). Thus, the viruses may be useful in the future as insertional mutagens. Their particular utility would be their ability to "tag" a gene: once the insertion was generated, the virus would provide a hybridization probe for the DNA near the target gene. Unfortunately, frequencies of target insertion may well prove to be generally less than those of spontaneous mutation; insertions into functionally diploid genes may well be undetectable.

Insertion events can also occur in the germ line of animals as well as in tissue culture. Inherited viruses are capable of provirus copy number amplification in vivo (Steffen et al. 1980, 1982; Herr and Gilbert 1982; Mowat and Bernstein 1983); reinfection of germ cells in adult animals presumably allows for the creation of new, inherited proviruses in the offspring. In a few cases, this process has been shown to be responsible for the creation of inherited mutations. A common allele of the *dilute* locus was apparently generated by the insertion of a provirus (Jenkins et al. 1981), and other coat color mutations known for many years (particularly in those genes known to be somatically mutable and contributing to variegated coat colors) may also be due to similar events. In a series of germ line integration events, it was shown that a developmentally regulated, recessive lethal mutation created by infection of embryos with M-MuLV was due to the insertion of the provirus into the collagen alpha 1 gene (Schnieke et al. 1983).

Finally, the most important of the mutations induced by the MuLVs must be described – the mutations which cause leukemia. It is now clear that at least in some tumors the provirus of M-MuLV lies very close to the known oncogene *c-myc*. We can probably assume that the insertion was the primary event leading to transformation and that in these tumors the virus acted to induce a high or aberrant expression of the *c-myc* gene product, ultimately resulting in some alteration in cell growth. The result is probably not complete cell transformation. If it were, and if proviral insertions were even approximately randomly distributed, one should be able to induce transformation in one cell by infection of only 10^6–10^7 cells. This has not been observed, and thus it is likely that secondary events must occur that are essential for the complete expression of the transformed phenotype.

It may be that replicating viruses are not absolutely necessary for such gene activations, although it is likely that the frequency is enormously greater when the virus is actively spreading in an animal. Recently, an oncogenic activation of the *c-mos* gene was reported (RECHAVI et al. 1982) to have occurred by the movement of Intracistemal A-Type Particle (IAP) sequences, endogenous DNAs closely related to the retroviruses in structure and probably in evolutionary history. Thus, intracellular activation of retroviruslike sequences can result in mutation.

The range of genetic events catalyzed by the retroviruses is clearly large. Essentially every event known to be catalyzed by transposable elements of bacteria and eukaryotes (gene disruption, gene activation, deletions, inversions) is mediated by the retroviruses. Thus, these genetic elements are likely to be extremely useful as tools for the molecular biologist. Ultimately, the MuLVs may become the agents of choice for mutagenesis, strain constrution, and gene transfer into mammalian cells in culture as well as into the germ lines of experimental animals.

Acknowledgments. The author was supported by a grant from the National Cancer Institute (NCI RO1 CA30488-03) and young investigator awards from the Searle Scholars Program and the Irma T. Hirschl Trust.

References

Aaronson SA, Barbacid M (1980) Viral genes involved in leukemogenesis: I. Generation of recombinants between oncogenic and nononcogenic mouse type-C viruses in tissue culture. J Exp Med 151:467–480

Abelson HT, Rabstein LS (1970) Lymphosarcoma: virus-induced thymic-independent disease in mice. Cancer Res 30:2213–2222

Battula N, Loeb LA (1979) The infidelity of avian myeloblastosis virus deoxyribonucleic acid polymerase in polynucleotide replication. J Biol Chem 249:4086–4093

Besmer P, Baltimore D (1977) Mechanism of restriction of ecotropic and xenotropic murine leukemia viruses and formation of pseudotypes between the two viruses. J Virol 21:965–973

Besmer P, Fan H, Paskind M, Baltimore D (1979) Isolation and characterization of a mouse cell line containing a defective moloney murine leukemia virus genome. J Virol 29:1023–1034

Blair DG, Oskarsson M, Wood TG, McClements WL, Fischinger PJ, Vande Woude GG (1981) Activation of the transforming potential of a normal cell sequence: a molecular model for oncogenesis. Science 212:941–943

Boone LR, Myer FE, Yang DM, Ou C-Y, Koh CK, Roberson LE, Tennant RW, Yang WK (1983) Reversal of Fv-1 Host range by in vitro restriction endonuclease fragment exchange between molecular clones of N-tropic and B-tropic murine leukemia virus genomes. J Virol 48:110–119

Chatis PA, Holland CA, Hartley JW, Rowe WP, Hopkins N (1983) Role for the 3′ end of the genome in determining disease specificity of Friend and Moloney murine leukemia viruses. Proc Natl Acad Sci USA 80:4408–4411

Cloyd MW, Hartley JW, Rowe WP (1980) Lymphomagenicity of recombinant Mink cell focus-inducing murine leukemia viruses. J Exp Med 151:542–552

Collins JK, Chesebro B (1981) Replication-defective Friend murine leukemia virus particles containing uncleaved *gag* polyproteins and decreased levels of envelope glycoprotein. J Virol 37:161–170

Crawford S, Goff SP (1984) Mutations in *gag* proteins P12 and P15 of Moloney murine leukemia virus block early stages of Infection. J Virol 49:909–917

Demsey A, Kawka D, Galuska S, Stackpole CW (1979) Assembly of a temperature-sensitive mutant of Rauscher murine leukemia virus at the cell surface induced by low temperature and by ligands. Virology 95:235–240

DesGroseillers L, Rassart E, Jolicoeur P (1983) Thymotropism of murine leukemia virus is conferred by its long terminal repeat. Proc Natl Acad Sci USA 80:4203–4207

Faller DV, Hopkins N (1977) RNAse T1-resistant oligonucleotides of an N- and a B-tropic murine leukemia virus of BALB/c: Evidence for recombination between these viruses. J Virol 24:609–617

Faller DV, Hopkins N (1978) T1 oligonucleotides that segregate with tropism and with properties of gp70 in recombinants between N- and B-tropic murine leukemia viruses. J Virol 26:153–158

Famulari NG (1983) Murine leukemia viruses with recombinant *env* genes: a discussion of their role in leukemogenesis. In: Cooper et al. (eds) Current Topics in Microbiology and Immunology, vol 103. Springer, Berlin Heidelberg New York, pp 75–108

Fan H, Chute H, Chao E, Feuerman M (1983) Construction and characterization of Moloney murine leukemia virus mutants unable to synthesize glycosylated *gag* polyprotein. Proc Natl Acad Sci USA 80:5965–5969

Fischinger PJ, Nomura S, Bolognesi DP (1975) A novel murine oncornavirus with dual ecotropic and xenotropic properties. Proc Natl Acad Sci USA 72:5150–5155

Gazdar AF, Oie A, Lalley P, Mass WW, Minna JD (1977) Identification of mouse chromosomes required for murine leukemia virus replication. Cell 11:949–956

Gerwin BI, Rein A, Levin JG, Bassin RH, Benjers BM, Kashmiri SVS, Hopkins D, O'Neill BJ (1979) Mutant of B-tropic murine leukemia virus synthesizing an altered polymerase molecule. J Virol 31:741–751

Gilboa E, Mitra SW, Goff S, Baltimore D (1979) A detailed model of reverse transcription and tests of crucial aspects. Cell 18:93–100

Goff SP, Gilboa E, Witte ON, Baltimore D (1980) Structure of the Abelson murine leukemia virus genome and the homologous cellular gene: studies with cloned viral DNA. Cell 22:777–785

Goff S, Traktman P, Baltimore D (1981) Isolation and properties of Moloney murine leukemia virus mutants: Use of a rapid assay for release of virion reverse transcriptase. J Virol 38:239–248

Goff SP, Tabin CJ, Wang JY-J, Weinberg R, Baltimore D (1982) Transfection of fibroblasts by cloned Abelson murine leukemia virus DNA and recovery of transmissible virus by recombination with Helper virus. J Virol 41:271–285

Goldfarb MP, Weinberg RA (1981) Generation of novel, biologically active Harvey sarcoma virus via apparent illegitimate recombination. J Virol 38:136–150

Gross L (1970) Oncogenic viruses, 2nd ed. Pergamon, New York

Haas M (1978) Leukemogenic activity of thymotropic, ecotropic, and xenotropic radiation leukemia virus isolates. J Virol 25:705–709

Hartley JW, Rowe WP (1976) Naturally occurring murine leukemia viruses in wild mice: Characterization of a new "amphotropic" class. J Virol 19:19–25

Hartley JW, Wolford NK, Old LJ, Rowe WP (1977) A new class of murine leukemia virus associated with development of spontaneous lymphomas. Proc Natl Acad Sci USA 74:789–792

Herr W, Gilbert W (1982) Germ line murine leukemia virus reintegration in AKR/J mice. Nature 296:865–868

Hopkins N, Jolicoeur P (1975) Variants of N-tropic leukemia virus derived from BALB/c mice. J Virol 16:991–999

Hopkins N, Traktman P, Whalen K (1976) N-tropic variants obtained after co-infection with N- and B-tropic murine leukemia viruses. J Virol 18:324–331

Hopkins N, Schindler J, Hynes R (1977a) Six NB-tropic murine leukemia viruses derived from a B-tropic virus of BALB/c have altered P30. J Virol 21:309–318

Hopkins N, Schindler J, Gottlieb PD (1977b) Evidence for recombination between N- and B-tropic murine leukemia viruses. J Virol 21:1074–1078

Huang AS, Besmer P, Chu L, Baltimore D (1973) Growth of pseudotypes of vesicular stomatitis virus with N-tropic murine leukemia virus coats in cells resistant to N-tropic virus. J Virol 12:659–662

Ishimoto A, Hartley JW, Rowe WP (1977) Detection and quantitation of phenotypically mixed viruses: Mixing of ecotropic and xenotropic murine leukemia viruses. Virology 81:263–269

Jenkins NA, Copeland NG, Taylor BA, Kee BK (1981) Dilute (*d*) coat color mutation of DBA/2J mice is associated with the site of integration of an ecotropic MuLV genome. Nature 293:370–374

Jolicoeur P (1979) The Fv-1 gene of the mouse and its control of murine leukemia virus replication. In: Cooper, et al (eds) Current Topics Microbiology and Immunology, vol 86. Springer, Berlin Heidelberg New York, pp 67–122

Jolicoeur P, Baltimore D (1976) Effect of FV-1 gene product on proviral DNA formation and integration in cells infected with murine leukemia viruses. Proc Natl Acad Sci USA 73:2236–2240

Jolicoeur P, Rosenberg N, Cotellessa A, Baltimore D (1978) Leukemogenicity of clonal isolates of murine leukemia viruses. J Natl Cancer Inst 60:1473–1476

Junghans RP, Boone LR, Skalka AM (1982) Products of reverse transcription in avian retrovirus analyzed by electron microscopy. J Virol 43:544–554

Khan AS, Martin MA (1983) Endogenous murine leukemia proviral long terminal repeats contain a unique 190-base-pair insert. Proc Natl Acad Sci USA 80:2699–2703

Lai M-HT, Verma IM, Tronick SR, Aaronson SA (1978) Mammalian retrovirus-associated RNase H is virus coded. J Virol 27:823–825

Landen B, Fox CF (1980) Isolation of BPgp70, a fibroblast receptor for the envelope antigen of Rauscher murine leukemia virus. Proc Natl Acad Sci USA 77:4988–4992

Levin JG, Seidman JG (1981) Effect of polymerase mutations on packaging of primer $tRNA^{pro}$ during murine leukemia virus assembly. J Virol 38:403–408

Levy JA (1979) Xenotropic type C viruses. In: Cooper et al (eds) Current Topics in Microbiology and Immunology, vol 79. Springer, Berlin Heidelberg New York, pp 109–213

Lilly F, Pincus T (1973) Genetic control of murine viral leukemogenesis. Adv Cancer Res 17:231–277

Linial M, Medeiros E, Hayward WS (1978) An avian oncovirus mutant (SE21Q1b) deficient in genomic RNA: Biological and biochemical characterization. Cell 15:1371–1381

Lu AH, Soong MM, Wong PKY (1979) Maturation of Moloney murine leukemia virus. Virology 93:269–274

Lueders K, Leder A, Leder P, Kuff E (1982) Association between a transposed alpha-globin pseudogene and retrovirus-like elements in the BALB/c mouse genome. Nature 295:426–428

Lung ML, Hering C, Hartley JW, Rowe WP, Hopkins N (1980) Analysis of the genomes of mink cell focus-inducing murine type-C viruses: a progress report. Cold Spring Harbor Symp Quant Biol 44:1269–1274

Manly KF, Buffett RF (1979) Differences in pathenogenicity among clonal sublines of a murine leukemia virus. J Virol 30:232–240

Mann R, Mulligan RC, Baltimore D (1983) Construction of a retrovirus packaging mutant and its use to produce Helper-free defective retrovirus. Cell 33:153–159

McCarter JA (1977) Genetic studies of the ploidy of Moloney murine leukemia virus. J Virol 22:9–15

McCarter JA, Ball JK, Frei JV (1977) Lower limb paralysis induced in mice by a temperature-sensitive mutant of Moloney leukemia virus. J Natl Cancer Inst 59:179–183

Mowat M, Berstein A (1983) Linkage of the Fv-2 gene to a newly reinserted ecotropic retrovirus in Fv-2 congenic mice. J Virol 47:471–477

Nowinski RC, Hays EF, Doyle T, Linkhart S, Medeiros E, Pickering R (1977) Oncornaviruses produced by murine leukemia cells in culture. Virology 81:363–370

Nusse R, Varmus HE (1982) Many tumors induced by the mouse mammary tumor virus contain a provirus integrated in the same region of the host genome. Cell 31:99–109

O'Donnell PV, Stockert E (1976) Induction of GIX antigen and Gross cell surface antigen after infection by ecotropic and xenotropic murine leukemia viruses in vitro. J Virol 20:545–554

Oie HK, Gazdar AF, Lalley PA, Russell EK, Minna JD, Delarco J, Todaro GJ (1978) Mouse chromosome 5 codes for ecotropic murine leukemia virus cell-surface receptor. Nature 274:60–62

Poiesz BJ, Ruscetti FW, Reitz MS, Kalyanaraman VS, Gallo RC (1981) (1981) Isolation of a new type of C retrovirus (HTLV) in primary uncultured cells of a patient with Sezary T-cell leukemia. Nature 294:268–271

Rapp UR, Nowinski RC (1976) Endogenous ecotropic mouse type C viruses deficient in replication and production of XC plaques. J Virol 18:411–417

Rapp UR, Todaro GJ (1978) Generation of new mouse sarcoma viruses in cell culture. Science 201:821–824

Rasheed S, Gardner MB, Chan E (1976) Amphotropic host range of naturally occurring wild mouse leukemia viruses. J Virol 19:13–18

Rechavi G, Givol D, Canaani E (1982) Activation of a cellular oncogene by DNA rearrangement: possible involvement of an IS-like element. Nature 300:607–611

Rein A (1982) Interference grouping of murine leukemia viruses: a distinct receptor for the MCF recombinant viruses in mouse cells. Virology 120:251–257

Rein A, Bassin RH (1978) Replication-defective ecotropic murine leukemia viruses: Detection and quantitation of infectivity using helper-dependent XC plaque formation. J Virol 28:656–660

Rein A, Kashmiri SVS, Bassin RH, Gerwin BI, Duran-Troise G (1976) Phenotypic mixing between

N- and B-tropic murine leukemia viruses: Infectious particles with dual sensitivity to Fv-1 restriction. Cell 7:373–379

Rein A, Gerwin BI, Bassin RH, Schwarm L, Schildlovsky G (1978) A replication-defective variant of Moloney murine leukemia virus: I. Biological characterization. J Virol 25:146–156

Rein A, Athan E, Benjers BM, Bassin RH, Gerwin BI, Slocum DR (1979a) Isolation of a replication-defective murine leukemia virus from cultured AKR leukemia cells. Nature 282:753–754

Rein A, Benjers BM, Gerwin BI, Bassin RH, Slocum DR (1979b) Rescue and transmission of a replication-defective variant of Moloney murine leukemia virus. J Virol 29:494–500

Repaske R, O'Neill RR, Steele PE, Martin MA (1983) Characterization and partial nucleotide sequence of endogenous type C retrovirus segments in human chromosomal DNA. Proc Natl Acad Sci USA 80:678–682

Reynolds RK, Stephenson JR (1977) Intracistronic mapping of the murine type C viral *gag* gene by use of conditional lethal replication mutants. Virology 81:328–340

Ruddle NH, Conta BS, Leinwald L, Kozak C, Ruddle F, Besmer P, Baltimore D (1978) Assignment of the receptor for ecotropic murine leukemia virus to mouse chromosome 5. J Exp Med 148:451–465

Ruta M, Murray MJ, Webb MC, Kabat D (1979) A murine leukemia virus mutant with a temperature-sensitive defect in membrane glycoprotein synthesis. Cell 16:77–88

Sacks TL, Devare SG, Blennerhassett GT, Stephenson JR (1978) Nonconditional replication mutants of type C and type D retroviruses defective in *Gag* gene-coded polyprotein post-translational processing. Virology 91:352–363

Schindler J, Hynes R, Hopkins N (1977) Evidence for recombination between N- and B-tropic murine leukemia viruses: Analysis of three virion proteins by sodium dodecyl sulfate-polyacrylamide gel electrophoresis. J Virol 23:700–707

Schnieke A, Harbers K, Jaenisch R (1983) Embryonic lethal mutation in mice induced by retrovirus insertion into the alpha 1(I) collagen gene. Nature 304:315–320

Schwartzberg P, Colicelli J, Goff SP (1983) Deletion mutants of Moloney murine leukemia virus which lack glycosylated *gag* protein are replication-competent. J Virol 46:538–546

Schwartzberg P, Colicelli J, Gordon ML, Goff SP (1984) Mutations in the *gag* gene of Moloney Murine leukemia virus: effects on production of virions and reverse transcriptase. J Virol 49:918–924

Shields A, Witte ON, Rothenberg E, Baltimore D (1978) High frequency of aberrant expression of Moloney murine leukemia virus in clonal infections. Cell 14:601–609

Shinnick TM, Lerner RA, Sutcliffe JG (1981) Nucleotide sequence of Moloney murine leukemia virus. Nature 293:543–548

Shoemaker C, Goff S, Gilboa E, Paskind M, Mitra SW, Baltimore D (1980) Structure of a cloned circular Moloney murine leukemia virus DNA molecule containing an inverted segment: Implications for retrovirus integration. Proc Natl Acad Sci USA 77:3932–3936

Shoemaker C, Goff S, Gilboa E, Paskind M, Mitra SW, Baltimore D (1981a) Structure of cloned retroviral circular DNAs: Implications for virus integration. Cold Spring Harbor Symp Quant Biol 45:711–717

Shoemaker C, Hoffmann J, Goff SP, Baltimore D (1981b) Intramolecular integration within Moloney murine leukemia virus DNA. J Virol 40:164–172

Steffen D, Weinberg RA (1978) The integrated genome of murine leukemia virus. Cell 15:1003–1010

Steffen DL, Bird S, Weinberg RA (1980) Evidence for the Asiatic origin of endogenous AKR-type murine leukemia proviruses. J Virol 35:824–835

Steffen DL, Taylor BA, Weinberg RA (1982) Continuing germ line integration of AKV proviruses during the breeding of AKR mice and derivative recombinant inbred strains. J Virol 42:165–175

Stephenson JR, Aaronson SA (1973) Characterization of temperature-sensitive mutants of murine leukemia virus. Virology 54:53–59

Stephenson JR, Aaronson SA (1974) Temperature-sensitive mutants of murine leukemia virus: III. Mutants defective in helper functions for sarcoma virus fixation. Virology 58:294–297

Stephenson JR, Reynolds RK, Aaronson SA (1972) Isolation of temperature-sensitive mutants of murine leukemia virus. Virology 48:749–756

Stephenson JR, Tronick SR, Aaronson SA (1974a) Temperature-sensitive mutants of murine leukemia virus: IV. Further characterization and evidence for genetic recombination. J Virol 14:918–923

Stephenson JR, Anderson GR, Tronick SR, Aaronson SA (1974b) Evidence for genetic recombination between endogenous and exogenous mouse RNA type C viruses. Cell 2:87–94

Stephenson JR, Tronick SR, Aaronson SA (1975) Murine leukemia virus mutants with temperature-sensitive defects in precursor polyepetide cleavage. Cell 6:543–548

Traktman P, Baltimore D (1982) Protease bypass of temperature-sensitive murine leukemia virus maturation mutants. J Virol 44:1039–1046

Tronick SR, Stephenson JR, Verma IM, Aaronson SA (1975) Thermolabile reverse transcriptase of a mammalian leukemia virus mutant temperature sensitive in its replication and sarcoma virus helper functions. J Virol 16:1476–1482

Van De Ven WJM, Van Zaane D, Onnekink C, Bloemers HPJ (1978) Impaired processing of precursor polypeptides of temperature-sensitive mutants of Rauscher murine leukemia virus. J Virol 25:553–561

Varmus HE (1982) Form and function of retroviral proviruses. Science 216:812–820

Varmus HE, Quintrell N, Ortiz S (1981) Retroviruses as mutagens: Insertion and excision of a nontransforming provirus alter expression of a resident transforming provirus. Cell 25:23–26

Verma IM (1975) Studies on reverse transcriptase of RNA tumor viruses: I. Localization of thermolabile DNA polymerase and RNAse H activities on one polypeptide. J Virol 15:121–126

Wang JYJ, Ledley F, Goff S, Lea R, Groner Y, Baltimore D (1984) The Mouse *c-abl* locus: Molecular Cloning and Characterization. Cell 36:349–356

Weiss RA, Wong AL (1977) Phenotypic mixing between avian and mammalian RNA tumor viruses: I. Envelope pseudotypes of Rous sarcoma virus. Virology 76:826–834

Witte ON, Baltimore D (1977) Mechanism of formation of pseudotypes between vesicular stomatitis virus and murine leukemia virus. Cell 11:505–511

Witte ON, Baltimore D (1978) Relationship of retrovirus polyprotein cleavages to virion maturation studied with temperature-sensitive murine leukemia virus mutants. J Virol 26:750–761

Wong PKY, McCarter JA (1973) Genetic studies of temperature-sensitive mutants of Moloney murine leukemia virus. Virology 53:319–326

Wong PKY, McCarter JA (1974) Studies of two temperature-sensitive mutants of Moloney murine leukemia virus. Virology 58:396–408

Wong PKY, MacLeod R (1975) Studies on the budding process of a temperature-sensitive mutant of murine leukemia virus with a scanning electron microscope. J Virol 16:434–442

Wong PKY, Gallick GE (1978) Preliminary characterization of a temperature-sensitive mutant of Moloney murine leukemia virus that produces particles at the restrictive temperature. J Virol 25:187–192

Wong PKY, Russ LJ, McCarter JA (1973) Rapid, selective procedure for isolation of spontaneous temperature-sensitive mutants of Moloney leukemia virus. Virology 51:424–431

Wyke JA, Bell JG, Beamand JA (1975) Genetic recombination among temperature-sensitive mutants of Rous sarcoma virus. Cold Spring Harbor Symp Quant Biol 39:897–905

Yeger H, Kalnins VI (1978) Immunocytochemical localization of gp70 over virus-related submembrane densities in *ts* mutant Rauscher murine leukemia virus-infected cells at the nonpermissive temperature. Virology 91:489–492

Yeger H, Kalnins VI, Stephenson JR (1976) Electron microscopy of mammalian type-C RNA viruses: Use of conditional lethal mutants in studies of virion maturation and assembly. Virology 74:459–469

Yeger H, Kalnins VI, Stephenson JR (1978) Type-C retrovirus maturation and assembly: Post-translational cleavage of the *gag*-gene coded precursor polypeptide occurs at the cell membrane. Virology 89:34–44

Yoshimura FK, Yamamura JM (1981) Four Moloney murine leukemia virus-infected rat cell clones producing replication-defective particles: Protein and nucleic acid analyses. J Virol 38:895–905

Yoshinaka Y, Ishigame K, Ohno T, Kageyama S, Shibata K, Luftig RB (1980) Preparations enriched for "immature" murine leukemia virus particles that remain in tissue culture fluids are deficient in $Pr65^{gag}$ proteolytic activity. Virology 100:130–140

Yuen PH, Wong PKY (1977) A morphological study on the ultrastructure and assembly of murine leukemia virus using a temperature-sensitive mutant restricted in assembly. Virology 80:260–274

Zavada J (1972) Pseudotypes of VSV with the coat of murine leukemia and avian myelobastosis viruses. J Gen Virol 15:183–191

Expression of Cellular Oncogenes*

R. MÜLLER[1] and I.M. VERMA[2]

1	Introduction	74
2	Characteristics and Expression of Retroviral Oncogenes and Their Cellular Homologs	76
2.1	The *fos* Gene of Murine Osteosarcoma Viruses	76
2.1.1	Characteristics of v-*fos* and Its Cellular Homolog	76
2.1.2	c-*fos* Expression	77
2.1.2.1	Expression of c-*fos* During Development of the Mouse Placenta	77
2.1.2.2	Expression of c-*fos* in Mouse Extraembryonal Membranes	78
2.1.2.3	Expression of c-*fos* in Mouse Fetuses and Postnatal Tissues	80
2.1.2.4	Expression of c-*fos* in Human Tissues	81
2.2	The Oncogenes of Feline Sarcoma Viruses: *fes*, *fms* and *fgr*	81
2.2.1	Characteristics and Expression of v-*fes*, v-*fms*, and v-*fgr*	81
2.2.2	Expression of c-*fms*	82
2.3	The *abl* Gene of Abelson Murine Leukemia Virus	83
2.3.1	Characteristics of v-*abl* and Its Cellular Homolog	83
2.3.2	Expression of c-*abl*	84
2.4	The Oncogenes of Harvey, Kirsten, Rasheed, and Balb Sarcoma Viruses: A Multi-Gene Family of *ras* Genes	85
2.4.1	The v-H-*ras* and v-K-*ras* Genes	85
2.4.2	Expression of c-H-*ras* and c-K-*ras*	87
2.5	Two Oncogenes of Avian Erythroblastosis Virus: *erb*-A and *erb*-B	88
2.5.1	Characteristics and Expression of v-*erb* Genes	88
2.5.2	Expression of c-*erb*	89
2.6	The *myb* Gene of Avian Myeloblastosis Type Viruses	89
2.6.1	Characteristics and Expression of v-*myb*	89
2.6.2	Expression of c-*myb*	90
2.7	The *myc* Gene of Avian Myelocytomatosis Viruses	91
2.7.1	Characteristics of v-*myc*	91
2.7.2	Expression of c-*myc*	91
2.8	The Oncogenes of Avian Sarcoma Viruses	93
2.8.1	Characteristics and Expression of v-*src*, v-*fps*, v-*yes*, and v-*ros*	93
2.8.2	Expression of c-*src*, c-*fps*, c-*yes*, and c-*ros*	94
2.9	The *rel* Gene of Avian Reticuloendotheliosis Virus	95
2.10	The c-*mos* Locus	95
2.11	The *sis* Gene of Simian Sarcoma Virus	97
3	Amplification, Rearrangement, and Translocation of c-*onc* Sequences in Tumor Cells	97
3.1	Gene Amplification and Rearrangements	97
3.2	Translocations	98
4	Mechanisms of Neoplastic Transformation: Role of c-*onc* Gene Products	100
4.1	Quantitative Modulations	100

* Literature survey for this review was terminated 1.9.83; Tables complemented 1.3.84

1 European Molecular Biology Laboratory, Postfach 102209, D-6900 Heidelberg

2 Molecular Biology and Virology Laboratory, The Salk Institute, P.O. Box 85800, San Diego, CA 92138, USA

Current Topics in Microbiology and Immunology, Vol. 112

4.2 Qualitative Alterations 101
4.3 Cell Type Specificity 102
5 Summary and Prospects 103
References 105

1 Introduction

Acutely oncogenic retroviruses induce a broad spectrum of malignant tumors (for review see GRAF and BEUG 1978; TEICH et al. 1982). They all contain genomic sequences that are responsible for the induction of neoplastic transformation in vivo and in vitro. These sequences, termed viral oncogenes (v-*onc*), originated from the normal cellular genome (BISHOP and VARMUS 1982). Cellular homologs (c-*onc*) of 20 different retroviral oncogenes have now been identified in a variety of species throughout the vertebrate phylum (Table 1) (STEHELIN et al. 1976; BISHOP and VARMUS 1982; BISHOP 1983), and in some instances even in invertebrates (SHILO and WEINBERG 1982). Another category of c-*onc* genes has been detected in the DNA of malignant cells by virtue of these genes' ability to induce neoplastic transformation when "transfected" into a mouse fibroblastic cell line in vitro (for review see COOPER 1982). In several cases, however, it has been shown that such sequences are in fact the cellular homologs of v-*onc* genes (DER et al. 1982; PARADA et al. 1982; SANTOS et al. 1982; MCCOY et al. 1983; SHIMIZU et al. 1983c).

In recent years, major research efforts have focused on elucidating the physiological role of c-*onc* sequences in normal cells. The ability of acutely oncogenic retroviruses to interfere with the control of normal cell proliferation and differentiation has led to the postulation that the cellular homologs of v-*onc* gene products may fulfill certain physiological functions in these processes (GRAF and BEUG 1978; ELLIS et al. 1981; BISHOP 1981, 1983; BISHOP and VARMUS 1982; GOYETTE et al. 1983). The advent of molecular cloning has made available an array of *onc*-specific probes which have been used to investigate the expression of c-*onc* genes in various tissues and cells from chickens (CHEN 1980; SHIBUYA et al. 1982; VENNSTRÖM and BISHOP 1982; GONDA et al. 1982a) and humans (WESTIN et al. 1982a, b; MÜLLER et al. 1983b). As part of a systematic study of the role of c-*onc* genes during pre- and postnatal development of the mouse, we have addressed two issues: (a) Are c-*onc* genes expressed during embryogenesis? (b) Is there differential expression of c-*onc* genes? As shown in Figure 1 several c-*onc* genes have indeed been found to be expressed during prenatal development and, furthermore, stage- and tissue-specific patterns of expression have been observed (MÜLLER et al. 1982, 1983a, c). These findings have lent support to the widely held notion that c-*onc* gene products participate in developmental processes.

The purpose of this article is to review our present knowledge of c-*onc* expression in normal vertebrate tissues, in normal differentiation processes, and during normal embryogenesis. We shall attempt to summarize the available information and propose mechanisms by which activation of c-*onc* genes may lead to neoplastic transformation.

Table 1. Properties of viral oncogenes[a]

Viral oncogene (v-*onc*)	Species of origin	Tumors induced	*onc* gene product: Size[b]	Intracellular location	Special properties
v-*src*	Chicken	Sarcoma	p60 *src*	Plasma membrane, cytoplasm	Tyrosine kinase
v-*fps*/v-*fes*	Chicken/cat	Sarcoma	p105–170 gag-*fps* p85-95 gag-*fes*	Plasma membrane, cytoplasm	Tyrosine kinase
v-*ros*	Chicken	Sarcoma	p68 gag-*ros*		
v-*yes*	Chicken	Sarcoma	p90 gag-*yes*	Plasma membrane, cytoplasm	Tyrosine kinase
v-*myb*, v-*ets*[c]	Chicken	Myeloblastosis erythroblastosis (myb-ets)	p48 env-*myb* p135 gag-*myb-ets*	Nucleus (*myb*)	
v-*myc*, v-*mil*[d]	Chicken	Myelocytomatosis, carcinoma, sarcoma, endothelioma	p90–200 gag-*myc* or gag-pol-*myc* or p100 gag-*mil* and p57 myc	Nuclear matrix (*myc*)	DNA-binding properties in vitro
v-*erb*-A	Chicken	?	p75 gag-*erb*-A	Cytoplasm	
v-*erb*-B	Chicken	Erythroblastosis, sarcoma	gp65 *erb*-B	Plasma membrane	Homology with EGF-receptor[f]
v-*ski*[e]	Chicken	?	p125 gag-*ski*	Nucleus	
v-*rel*	Turkey	Lymphatic leukemia	p58 *rel*	?	
v-*abl*	Mouse/cat	T- and B-cell leukemia	p90-160 gag-*abl*	Plasma membrane (inner surface)	Tyrosine kinase
v-*fos*	Mouse	Osteosarcoma	p55 *fos* p75 gag-*fos*	Nucleus	
v-*mos*	Mouse	Rhabdomyosarcoma	p37 env-*mos*	Cytoplasm	
v-*raf*[e]	Mouse	Fibrosarcoma	p90/p75 gag-*raf*	Cytoplasm	
v-H-*ras*/v-*bas* (v-*has*)	Rat/mouse	Erythroleukemia, sarcoma	p21 ras^{H} p29 gag-*ras*	Plasma membrane (inner surface)	GTP-binding properties
v-K-*ras* (v-*kis*)	Rat	Erythroleukemia	p21 ras^{K}	Plasma membrane (inner surface)	GTP-binding properties
v-*fms*	Cat	Sarcoma	gp120–180 gag-*fms*	Plasma membrane	Glycoprotein
v-*fgr*	Cat	Sarcoma	p70 gag-*fgr*	?	γ-Actin (N-terminus); Tyrosine kinase (C-terminus)[g]
v-*sis*	Monkey/cat	Sarcoma	p28 env-*sis*	?	Homology with PDGF

[a] Table complemented 1.3.84
[b] Size estimated from SDS-PAGE; in some instances different from size predicted by nucleotide sequence analysis (e.g., *fos, mos*)
[c] Second oncogene of E26 virus
[d] Second oncogene of MH2 virus; also termed *mht*, homologous to *raf*
[e] Oncogene of defective retrovirus generated in vitro (Stavnezer et al. 1981; Rapp et al. 1983a, b)
[f] Downward et al. 1984
[g] Naharro et al. 1984

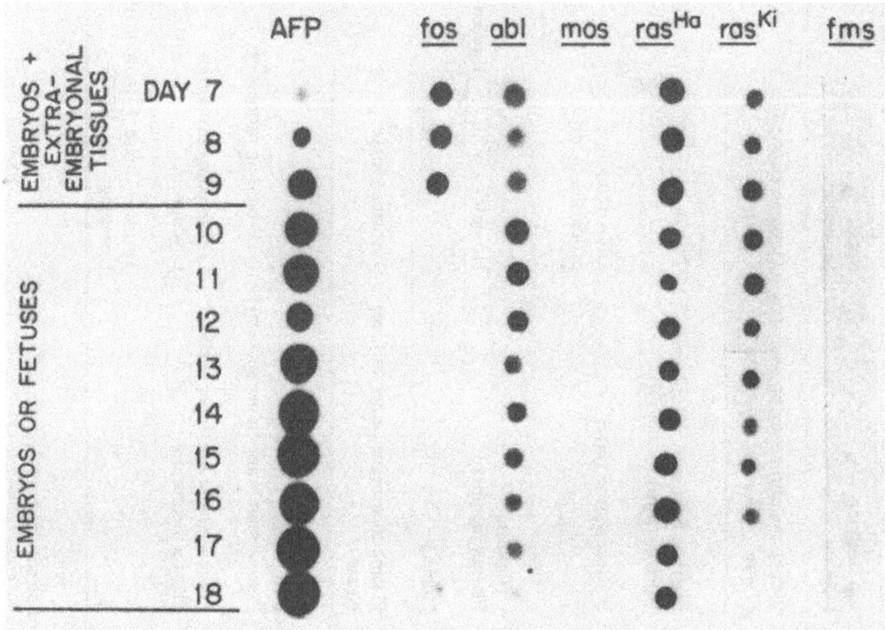

Fig. 1. Expression of cellular oncogenes during mouse prenatal development. Probes used were alpha-fetoprotein (AFP, obtained from S. Tilghman; included as a control); *fos* (FBJ murine osteosarcoma virus oncogene); *abl* (Abelson leukemia virus oncogene, obtained from S. Goff and David Baltimore); *mos* (moloney sarcoma virus oncogene ras^{Ha}, ras^{Ki} (Harvey and Kirsten MuSV oncogenes, obtained from E. Scolnick and colleagues); and *fms* (SM-FeSV oncogene, obtained from C. Sherr). (For experimental details see MÜLLER 1983 and MÜLLER et al. 1982, 1983c)

2 Characteristics and Expression of Retroviral Oncogenes and Their Cellular Homologs

2.1 The *fos* Gene of Murine Osteosarcoma Viruses

2.1.1 Characteristics of v-*fos* and Its Cellular Homolog

The FBJ murine osteosarcoma virus (FBJ-MSV) was isolated from a spontaneous osteosarcoma in a CFl mouse. When inoculated into newborn mice, FBJ-MSV induces specifically chondro-osseous neoplasms, which are usually associated with bones or the diaphragm. The tumors show locally invasive growth, generally do not give rise to metastases, and have been classified as low-grade malignant tumors (FINKEL et al. 1966; FINKEL and BISKIS 1968; WARD and YOUNG 1976). It has been suggested that periosteal cells may be the target cells for neoplastic transformation by FBJ-MSV (WARD and YOUNG 1968). FBJ-MSV also transforms fibroblasts in vitro (LEVY et al. 1973). A 55000-dalton (55 K) phosphoprotein encoded by the oncogene (v-*fos*) of FBJ-MSV has been identified (CURRAN and TEICH 1982; CURRAN et al. 1982). It has recently been found that FBR murine osteosarcoma virus (FBR-MSV), which originated from

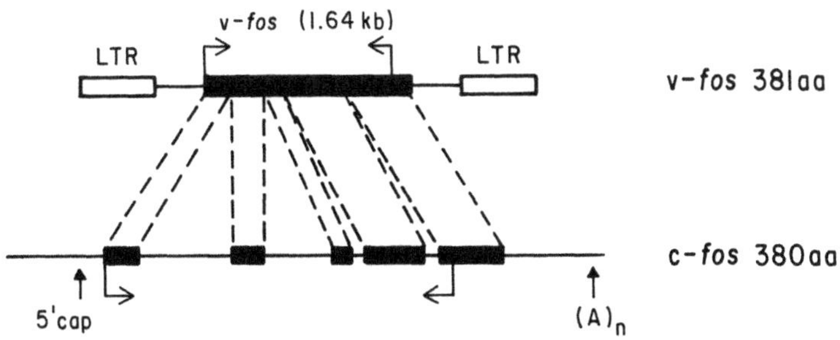

Fig. 2. Organization of fos gene. Structures derived by nucleotide sequence analyses of v-*fos* (FBJ-MSV) and c-*fos* (mouse) gene (VAN BEVEREN et al. 1983) Black boxes in c-*fos* represent exons. The *bent arrows* (↳) indicate the coding domain. *LTR*, long terminal repeat

a radiation-induced mouse osteosarcoma (FINKEL et al. 1975; LEE et al. 1979), also contains the v-*fos* oncogene. The transforming protein of FBR-MSV, however, appears to be expressed as a gag-*fos* fusion protein of 75 K (P75) (T. Curran, S. Enami, C. Van Beveren, and I.M. Verma, unpublished data). Cellular homologs (c-*fos*) of v-*fos* have been identified in the genomes of various vertebrates (CURRAN et al. 1982) as well as in *Drosophila* (I.M. Verma, unpublished data). Nucleotide sequence analyses revealed that both the mouse and human c-*fos* genes represent complete transcriptional units containing three intervening sequences (Fig. 2) (CURRAN et al. 1983; VAN BEVEREN et al. 1983; VAN STRAATEN et al. 1983). Although the *fos* genes are evolutionarily well conserved, the v-*fos* gene product in FBJ-MSV was found to lack 104 bp near its carboxy terminus when compared with its cellular progenitor. Consequently, v-*fos* and c-*fos* proteins have different carboxy termini (VAN BEVEREN et al. 1983; VAN STRAATEN et al. 1983). Both the v-*fos* and c-*fos* proteins migrate on sodium dodecyl sulfate polycrylamide gels (SDS-PAG) with a molecular weight of 55 K and have been localized to the cell nucleus (CURRAN et al. 1984). The c-*fos* (human) gene has been assigned to chromosome 2 (BARKER et al. 1983).

2.1.2 c-*fos* Expression

Expression of the c-*fos* gene has been extensively analyzed during development of the mouse fetus, placenta, and extraembryonal membranes (also referred to as fetal membranes; i.e., visceral yolk sac and amnion) as well as in a variety of postnatal mouse tissues. A diagrammatic representation of a midgestation conceptus [i.e., fetus plus all extraembryonal tissues (placenta and membranes)] is shown in Fig. 3. The results of these studies are summarized in the following sections.

2.1.2.1 Expression of c-fos During Development of the Mouse Placenta

c-*fos* gene transcripts were observed at relatively high, but nearly constant, levels in the undissected placenta throughout the gestational period and in 6th–

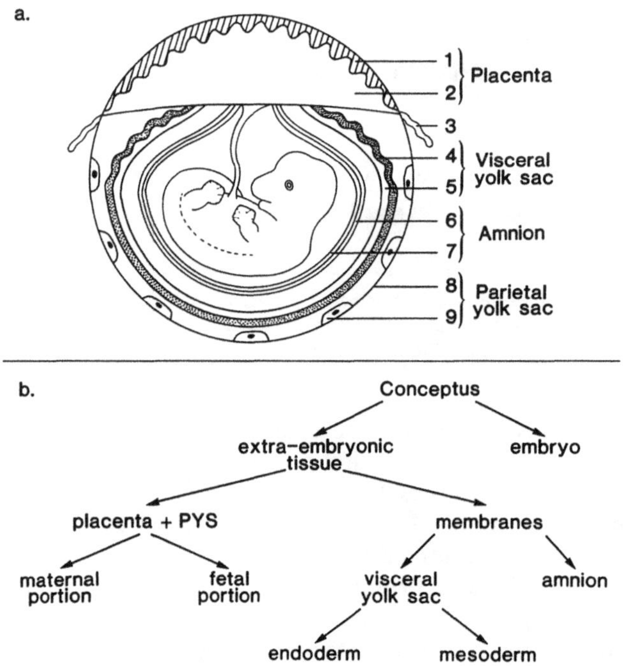

Fig. 3a, b. Dissection of midgestation conceptus. **a** Diagram of sectioned conceptus at day 12–18 of gestation. **b** Dissection strategy used for analysis of c-*onc* expression (Figs. 4, 5 and 8)

9th-day conceptuses, which consist largely of tissues participating in the formation of the placenta (Figs. 1, 4a) (MÜLLER et al. 1982, 1983a). Analysis of surgically dissected placentas, however, revealed that transcripts from the c-*fos* gene were approximately 15-fold more abundant in the outer portion of the midgestation placenta (primarily fetus-derived undifferentiated cytotrophoblast and maternal decidua) relative to the inner moiety (predominantly differentiated syncytiotrophoblast) (Fig. 4b) (MÜLLER et al. 1983a). This finding suggests that the majority of c-*fos* transcripts in the midgestation placenta originates either from the maternally derived decidua basalis or from the fetal cytotrophoblast. In this context, it should be noted that significant c-*fos* expression could also be detected in a trophoblast-like teratocarcinoma cell line (3TDM), but not in other embryonal carcinoma cells or embryoid bodies (Müller 1983; R. Müller and E.D. Adamson, unpubl. results). In the inner placenta, the level of c-*fos* transcripts was found to increase gradually as gestation proceeded, and to reach a level approximately 50% of that found in the outer moiety. This increase in c-*fos* expression may be correlated with the differentiation of chorionic tissue, which in the mouse is fused with the inner placenta (MÜLLER et al. 1983a).

2.1.2.2 Expression of c-fos in Mouse Extraembryonal Membranes

The level of c-*fos* transcripts was found to be low in 10th–12th-day extraembryonal membranes (MÜLLER et al. 1982, 1983a), but increased approximately

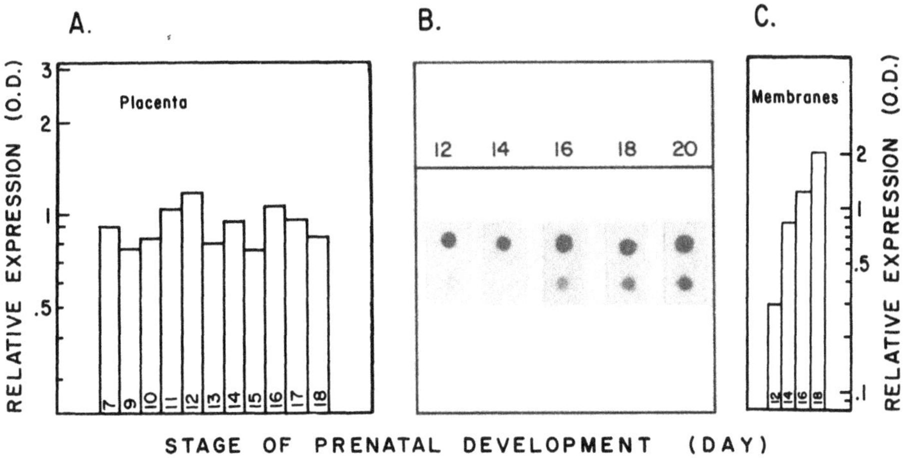

Fig. 4A–C. Quantification of the levels of *c-fos* transcripts during mouse prenatal development. **A** Placenta, days 7–18; **B** outer and inner placenta – the *upper lane* is outer placenta, the *lower lane* inner placenta; **C** extraembryonal membranes (including amnion and visceral yolk sac)

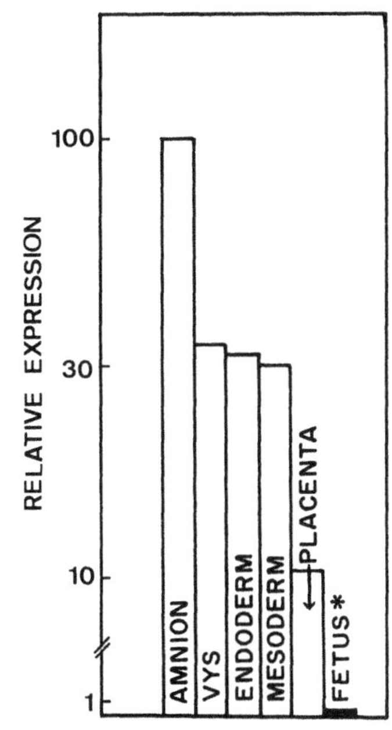

Fig. 5. Relative levels of *c-fos* transcripts in surgically separated extraembryonal tissues. * indicates barely detectable levels; *VYS*, visceral yolk sac

six fold between the 12th and 18th days of gestation to a level slightly higher than that observed in the placenta (Fig. 4c) (MÜLLER et al. 1983a). A detailed analysis of the microsurgically isolated components of the late-gestation extraembryonal membranes showed that the levels of c-*fos* transcripts were higher in the visceral yolk sac (VYS) endoderm and mesoderm than in the placenta. The highest level of expression of c-*fos* was observed in the amnion (Fig. 5) (MÜLLER et al. 1983a). The level of c-*fos* transcripts in the 18th-day amnion

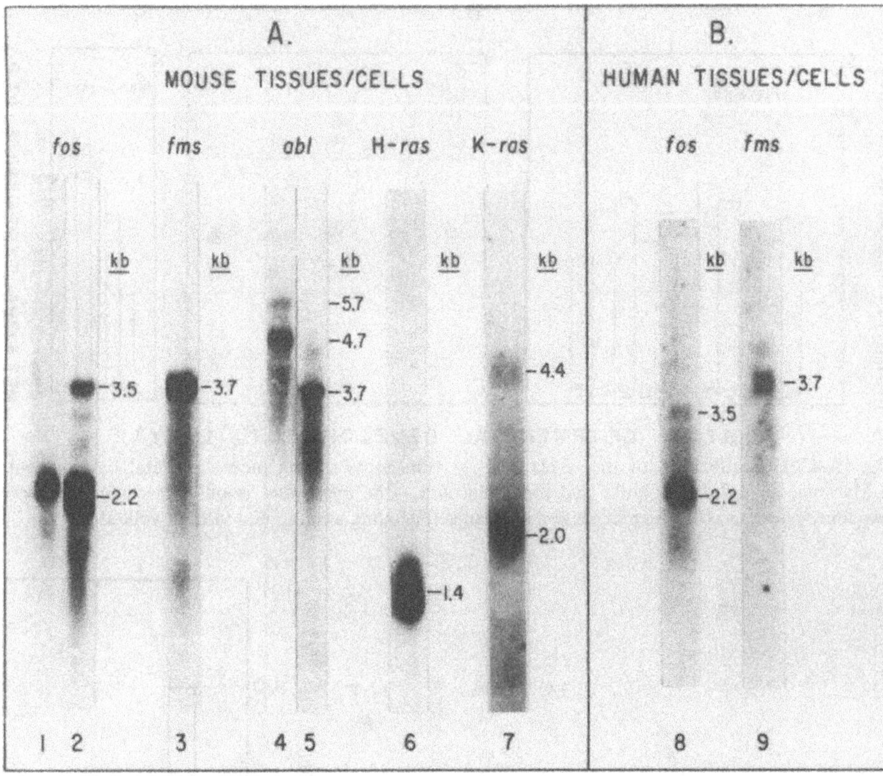

Fig. 6A, B. Size of *c-onc* transcripts. Northern blot analysis of poly(A^+) RNA from various tissues and cells of mouse **A** and human **B** origin. *1,* neonatal "bone"; *2* 17th-day membranes; *3* 16th-day placenta; *4* 11th-day embryo; *5* adult testes; *6* 12th-day embryo; *7* 12th-day membranes; *8* term placenta; *9* BeWo cells

was found to be close to the level of v-*fos* transcripts in cells transformed by FBJ-MSV. This finding is of particular interest, since it shows that a high level of c-*fos* expression is not sufficient to induce neoplastic transformation of amniotic cells (see section 4).

The size of the c-*fos* (mouse) gene transcript was determined to be 2.2 kb in all tissues and cells analyzed (Fig. 6A) (MÜLLER et al. 1982, 1983a). Sequence analysis of the c-*fos* gene would predict the c-*fos* RNA transcript to be about 2.2 kb following the removal of introns and termination at the putative poly (A) addition signal (VAN BEVEREN et al. 1983). In some instances, a second c-*fos*-related RNA species of approximately 3.5 kb could be identified (Fig. 6A, lane 2) (MÜLLER et al. 1983a), which presumably represents an unspliced mRNA precursor.

2.1.2.3 Expression of c-fos in Mouse Fetuses and Postnatal Tissues

Expression of c-*fos* was found to be low in fetuses throughout the gestational period from the 10th to the 17th day, but increased more than five fold at later stages of prenatal development (Fig. 1) (MÜLLER et al. 1982, 1983a). Analyses of a variety of tissues from newborn mice showed elevated levels of c-*fos* expression only in "bone" (rib cage and vertebra, including bone marrow, muscles, and part of the pleura) and "skin" tissue (including the subcutaneous connective tissues, muscles, and part of the peritoneum) (MÜLLER et al. 1982; MÜLLER 1983). This finding is intriguing since the *fos* sequences in FBJ-MSV have been implicated in the induction of osteosarcomas in newborn mice (FINKEL et al. 1966; FINKEL and BISKIS 1968; CURRAN and TEICH 1982; CURRAN et al. 1982).

Further analyses, however, have shown that expression of c-*fos* in postnatal tissues is more likely to originate from macrophages than from cells involved in the formation of connective tissues (e.g., bone, cartilage, muscle) (R. Müller, unpublished observation).

2.1.2.4 Expression of c-fos in Human Tissues

The results obtained from an analysis of c-*fos* expression in human tissues and cells closely resembled those described above for mouse tissues (MÜLLER et al. 1983b). The highest levels of c-*fos* transcripts were detected in human term chorion and amnion. Elevated concentrations of c-*fos* transcripts were also observed in term placenta and in a trophoblastic cell line (BeWo) derived from a human gestational choriocarcinoma relative to a number of other tissues and cells, including embryonal carcinoma cells, fetal fibroblasts, and a variety of postnatal tissues. The transcript from the c-*fos* (human) gene was found to be very similar in size to its murine counterpart (Fig. 6B). It thus appears that not only the structure of the c-*fos* gene but also the control mechanisms regulating its expression have been conserved during evolution.

2.2 The Oncogenes of Feline Sarcoma Viruses: *fes*, *fms*, and *fgr*

2.2.1 Characteristics and Expression of v-*fes*, v-*fms*, and v-*fgr*

Feline sarcoma viruses (FeSV) cause predominantly fibrosarcomas in cats (for review see HARDY 1980; TEICH et al. 1982). Both the Gardner-Arnstein (GA) and Snyder-Theilen (ST) strains of FeSV (GARDNER et al. 1970; SNYDER and THEILEN 1969) have acquired the same cat cellular sequences, denoted *fes* (Fig. 7A) (FRANKEL et al. 1979; FRANCHINI et al. 1981; COFFIN et al. 1981), which are homologous to the oncogene of Fujinami sarcoma virus (*fps*) (BARBACID et al. 1981; SHIBUYA et al. 1980, 1982; SHIBUYA and HANAFUSA 1982; HAMPE et al. 1982; see also section 2.8). The McDonough strain (SM) of FeSV (MCDONOUGH 1971) contains an *onc* gene which is distinct from that found in GA- or ST-FeSV and has been designated v-*fms* (Fig. 7) (COFFIN et al. 1981). The transforming proteins of these viruses are synthesized as *gag-onc* polyproteins and designated – $p85^{gag\text{-}fes}$ (ST-FeSV); $p95^{gag\text{-}fes}$ (Ga-FeSV), and $gp180^{gag\text{-}fms}$

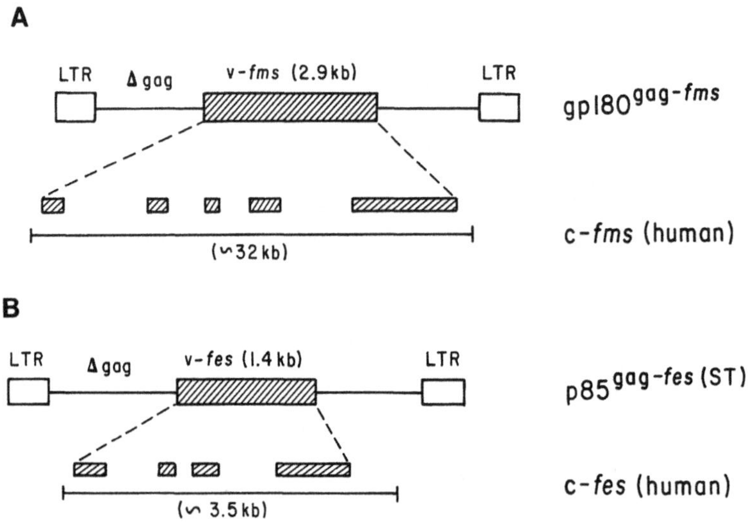

Fig. 7 A, B. Organization of *fms* and *fes* genes. **A** The structure of SM-FeSV proviral DNA (Donner et al. 1982) and of c-*fms*-specific sequences in human DNA (Heisterkamp et al. 1983) are shown. **B** The structure of ST-FESV proviral DNA containing the *fes* oncogene (Franchini et al. 1981) and the organization of *fes*-specific sequences in human DNA Groffen et al. 1982 are displayed. Hatched boxes represent exons

(SM-FeSV). SM-FeSV-transformed cells contain additional *fms*-related proteins of 140 K and 120 K (Barbacid et al. 1980b; Donner et al. 1982a; Anderson et al. 1982). All three forms of *fms*-related proteins are glycosylated. The cellular progenitors of the *fes* and *fms* oncogenes have been identified in the genome of normal cells (Frankel et al. 1979; Donner et al. 1982b). Recently, a third transforming gene (*fgr*) of feline sarcoma virus origin has been described. This oncogene is expressed as a 70 K gag-*fgr* fusion protein (Naharro et al. 1983).

c-*fes* gene expression has not yet been verified, although in one report a 92 K protein from normal feline cells has been identified by using antisera raised against ST-FeSV-transformed cells (Barbacid et al. 1981). c-*fes* transcripts have been identified in some human ovarian carcinomas, lung adenocarcinomas, and acute myelogenous leukemias (D. Slamon, personal communication).

The human cellular homologs of the *fes* and *fms* genes have been assigned to chromosome 15 (Dalla-Favera 1982; Heisterkamp et al. 1982) and chromosome 5 (P. Barker and F. Ruddle, personal communication), respectively.

2.2.2 Expression of c-*fms*

The expression pattern of the c-*fms* gene observed during mouse prenatal development bears some conspicuous similarities with that of the c-*fos* gene (Fig. 1) (Müller et al. 1983c). For instance, the level of c-*fms* transcripts was found

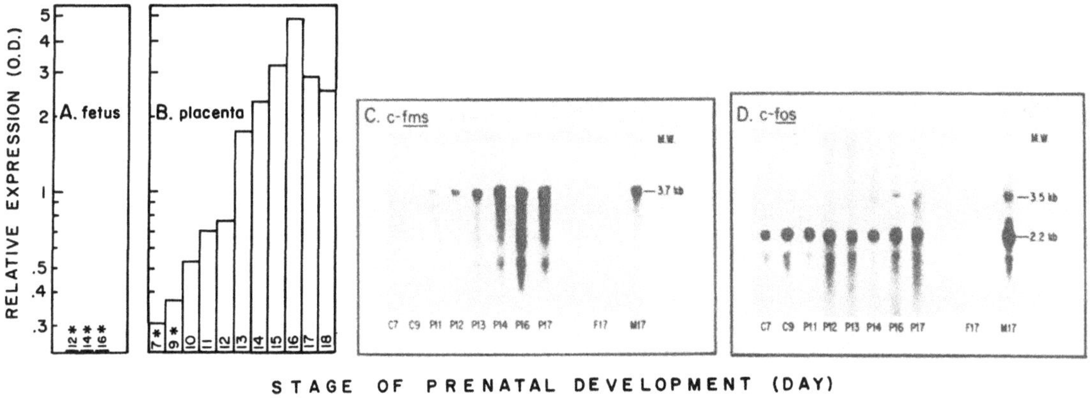

Fig. 8A–D. Comparison of c-*fms* and c-*fos* expression during prenatal development. Relative levels of c-*fms* transcripts in the fetus **A** and placenta **B** at various stages of prenatal development * indicates levels lower than 0.1. **C, D** Northern blot analysis of c-*fms* and c-*fos* transcripts from day-7 and -9 conceptuses (*C7, C9*), placenta from various stages of development (*P11–P17*), day-17 fetus (*F17*), and day 17 membranes (*M17*)

to be very low in fetuses at any developmental stage, but high in late-gestation placenta and extraembryonal membranes (Fig. 8A) (MÜLLER et al. 1983c). As is the case for c-*fos* expression, the concentration of transcripts from the c-*fms* gene gradually increased during development of the extraembryonal membranes. However, c-*fos* and c-*fms* are not expressed concordantly. In contrast to c-*fos,* the concentration of c-*fms* transcripts increased in the undissected placenta during mouse prenatal development (approximately 15-fold relative to 7th–9th-day conceptuses) before reaching a plateau at the 14th–16th day of gestation (Fig. 8b, c) (MÜLLER et al. 1983c). Moreover, the levels of c-*fms* transcripts were similar in the late-gestation amnion, VYS, and both moieties of the placenta (MÜLLER et al. 1983a). Figure 8c, d compares the stage-specific patterns of c-*fms* and c-*fos* expression. No significant expression of c-*fms* could be detected in any postnatal tissue (MÜLLER et al. 1983c). When human tissues and cells were analyzed, high levels of c-*fms* transcripts were found only in term placenta, fetal membranes, and in the trophoblastic BeWo cell line (MÜLLER et al. 1983b). Transcripts from both c-*fms* (mouse) and c-*fms* (human) were determined to be 3.7 kb (Fig. 6A, B) (MÜLLER et al. 1983b, c). As in the case of c-*fos,* the expression pattern of c-*fms* would be compatible with a specific function of its encoded protein in differentiation processes occurring during the development of mammalian extraembryonal tissues.

2.3 The *abl* Gene of Abelson Murine Leukemia Virus

2.3.1 Characteristics of v-*abl* and Its Cellular Homolog

Abelson murine leukemia virus (A-MuLV) induces lymphosarcomas of pre-B-cell origin (ABELSON and RABSTEIN 1970) and T-cell lymphomas (COOK 1982)

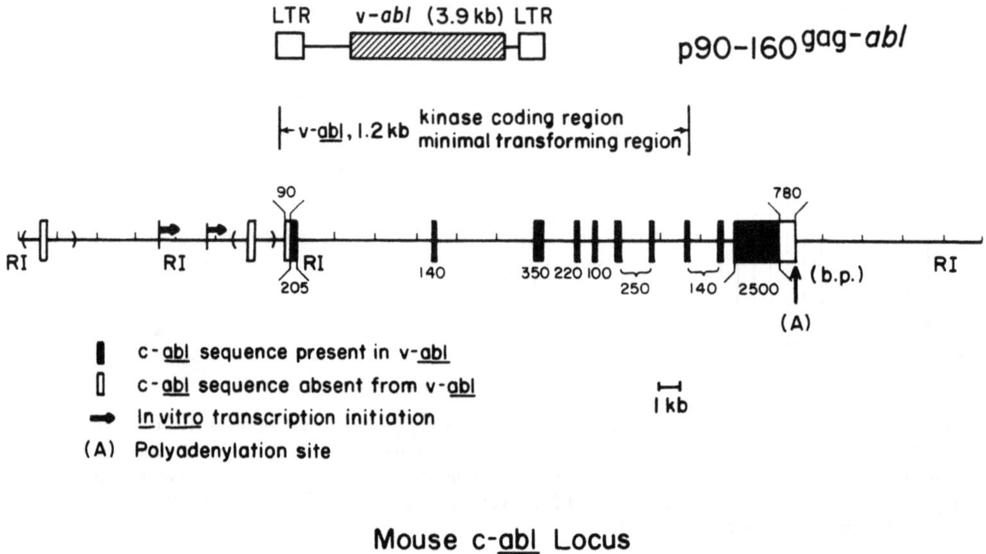

Fig. 9. Organization of *abl* genes. The structure of Abelson murine leukemia virus proviral DNA (Goff et al. 1980) and organization of the mouse c-*abl* locus. (Courtesy of J. Wang and D. Baltimore)

in mice and transforms fibroblasts and lymphocytes in vitro (Witte et al. 1978). The genome of A-MuLV contains an oncogene (v-*abl*) expressed as a gag-*abl* fusion protein of 160 K (P160) (Fig. 9) (Witte et al. 1978). The v-*abl* protein is associated with tyrosine protein kinase activity, which is believed to be instrumental in the transformation of both lymphocytes and fibroblasts (Reynolds et al. 1980; Witte et al. 1980a, b; Sefton et al. 1981). The v-*abl* protein appears to be located in the cytoplasmic membrane (Witte et al. 1979a). It has been suggested that P160 may have an affinity for membranes because it contains tightly bound lipid (Sefton et al. 1982). Cellular homologs (c-*abl*) of v-*abl* have been identified in a variety of vertebrate cells (Goff et al. 1980) and in *Drosophila* (Shilo and Weinberg 1980; Hoffman-Falk et al. 1983). The human c-*abl* sequences homologous to v-*abl* are distributed along 30 kb of human DNA (Heisterkamp et al. 1983b) and contains a series of small exons separated by long introns. The c-*abl* (mouse) locus is also about 30 kb and contains a large number of intervening sequences (J. Wang and D. Baltimore, personal communication). The c-*abl* (mouse) gene has been reported to be located on chromosome 2 (Goff et al. 1982), whereas c-*abl* (human) has been assigned to chromosome 9 (Heisterkamp et al. 1982).

2.3.2 Expression of c-*abl*

Expression of c-*abl* could be detected in mouse fetuses throughout prenatal development, but the level of transcripts gradually decreased between the 10th and 18th days of gestation (Fig. 1) (Müller et al. 1982). The concentration of c-*abl* transcripts was found to be approximately five fold greater on the 10th day than on the 18th day of gestation (Müller et al. 1982, unpublished data). Similarly, the level of c-*abl* transcripts was about 5-fold greater in midgestation fetuses than in extraembryonal tissues or 6th–9th-day conceptuses. It is interesting that A-MuLV induces lymphoid tumors and that the highest levels of c-*abl* expression may coincide with the development of the hematopoietic system of the mouse. Analysis of a variety of postnatal tissues revealed that c-*abl* is expressed at readily detectable levels in spleen, thymus, and thymic tumor cells (MÜLLER et al. 1982; MÜLLER 1983; WANG and BALTIMORE 1983). The highest level of c-*abl* transcripts, however, was found in the testes of adult mice, where it was more than five fold higher than in any other tissue analyzed (MÜLLER et al. 1982).

Two species of c-*abl* transcripts of approximately 4.7 kb and 5.7 kb were detected in all pre- and postnatal mouse tissues and cells analyzed (Fig. 6) (MÜLLER et al. 1982; MÜLLER 1983). The testes of adult mice, however, were found to contain an additional transcript of 3.7 kb, which was considerably more abundant than the two other mRNAs (Fig. 6A) (MÜLLER et al. 1982).

In normal cells from mouse lymphatic tissues, a protein of 150 K (NCP150) has been detected by immunoprecipitation using tumor regressor serum (WITTE et al. 1979b). This protein appears to be the cellular counterpart of the v-*abl* gene product (PONTICELLI et al. 1982). In contrast to the v-*abl*-encoded protein, however, no phosphotyrosine residues or protein kinase activity could be detected with NCP150 (PONTICELLI et al. 1982).

2.4 The Oncogenes of Harvey, Kirsten, Rasheed, and Balb Sarcoma Viruses: A Multi-Gene Family of *ras* Genes

2.4.1 The v-H-*ras* and v-K-*ras* Genes

Harvey murine sarcoma virus (HaSV), Kirsten murine sarcoma virus (KiSV), Rasheed (rat) sarcoma virus (RaSV), and Balb mouse sarcoma virus (Balb-MSV) cause erythroblastosis, extrathymic lymphomas, and sarcomas in rodents (HARVEY 1964; KIRSTEN and MAYER 1967; RASHEED et al. 1978; PETERS et al. 1974). Their oncogenes (v-*ras*) which originated from normal rat (HaSV, KiSV, and RaSV) or mouse (Balb-MSV) genomic sequences encode proteins that are biochemically, immunologically, and functionally related (SCOLNICK et al. 1979; SHIH TY et al. 1979; YOUNG et al. 1979, 1981; DEFEO et al. 1981; ELLIS et al. 1981). The oncogenes of RaSV and Balb-MSV (referred to as v-*bas*) have been shown to be more closely related to the transforming gene v-H-*ras* of HaSV than to the oncogene v-K-*ras* of KiSV (YOUNG et al. 1979, 1981; ANDERSEN et al. 1982; GONDA et al. 1982b). The transforming gene of RaSV is expressed as a gag-*ras* fusion protein of 29 K (P29) (YOUNG et al. 1979, 1981). In contrast, the *ras* genes of HaSV and KiSV encode 21 K proteins (P21) that do not contain helper virus sequences. P21 is bound to the inner surface of the cytoplasmic membrane (WILLINGHAM et al. 1980; SHIH et al. 1982). Viral P21 has been shown to autophosphorylate on threonine (SHIH TY et al. 1980) and to possess guano-

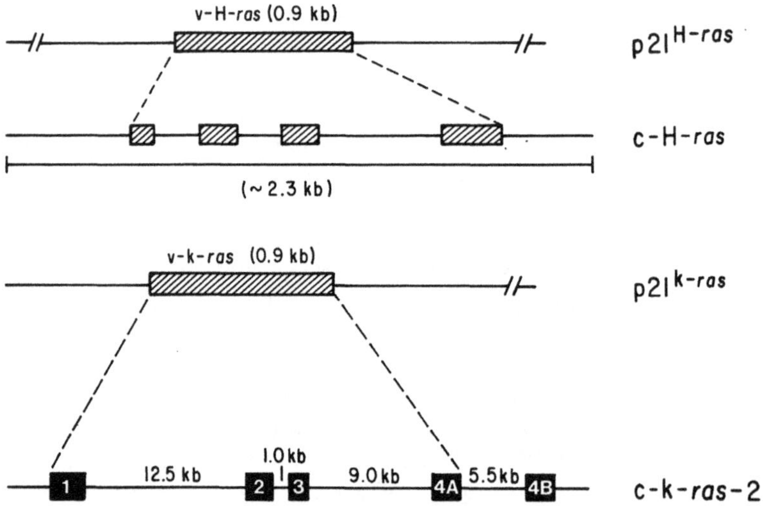

Fig. 10. Organization of *ras* genes. The structures of V-H-*ras* and its cellular homolog in the human genome (C-H-*ras*-1) (CHANG et al. 1982b) are shown. The structure of v-K-*ras* and its cellular counterpart in human DNA were derived by nucleotide sequence analyses. (MCGRATH et al. 1983; SHIMIZU et al. 1983b)

sine diphosphate (GDP)- and guanosine triphosphate (GTP)-binding activity (SCOLNICK et al. 1979).

Despite the kinship of the v-*onc*-encoded proteins of HaSV and KiSV, DNA probes specific for H-*ras* and K-*ras* sequences have been derived that hybridize with different restriction fragments of rat genomic DNA. This indicates that c-H-*ras* and c-K-*ras* represent different cellular genes within the *ras* gene family (ELLIS et al. 1981). The rat genome contains two distinct fragments, each carrying v-H-*ras*-specific sequences. As shown in Fig. 10, the c-H-*ras*-I (rat) contains three intervening sequences, whereas the c-H-*ras*-II (rat) fragment is devoid of introns and is colinear with v-H-*ras* (DHAR et al. 1982). [In an earlier publication, c-H-*ras*-I is without introns and c-H-*ras*-II contains introns (DEFEO et al. 1981). However, in recent papers the gene containing the introns is referred to as c-H-*ras*-I.] In contrast to the rat, the genome of the laboratory mouse (*Mus musculus*) contains only one fragment carrying H-*ras*-specific sequences (ELLIS et al. 1981). Another mouse strain, *Mus pahari,* however, has at least ten copies of c-H-*ras* sequences (CHATTOPADHYAY et al. 1982). The human genome contains two fragments homologous to v-H-*ras*. One of these fragments, c-H-*ras*-I (human), is structurally closely related to c-H-*ras*-I (rat), whereas the other fragment, c-H-*ras*-II (human), does not contain intervening sequences and is more divergent from v-H-*ras* (CHANG et al. 1982a). The c-H-*ras*-I (human) gene has been assigned to chromosome 11 (MCBRIDE et al. 1982; DE MARTINVILLE et al. 1983).

Sequences homologous to K-*ras* are also present as two copies in the genomes of rats (quoted in ELLIS et al. 1982) and humans (CHANG et al. 1982a).

The structure and organization of the c-K-*ras*-2 (human) gene has recently been elucidated (SHIMIZU et al. 1983a; MCGRATH et al. 1983; CAPON et al. 1983). It appears that c-K-*ras*-2 has four exons which constitute the v-K-*ras* coding domain (Fig. 10). Additional sequences from another exon (4b) are located in v-K-*ras* 3′ from the termination codon. A pseudogene of c-K-*ras* has also been identified. The c-K-*ras*-2 (human) sequences are located on chromosome 12 (SAKAGUCHI et al. 1983).

Recently, a third member of the *ras* gene family has been detected by DNA transfection experiments (HALL et al. 1983; MURRAY et al. 1983; SHIMIZU et al. 1983b). This gene has been termed N-*ras* and is located on human chromosome 1 (HALL et al. 1983).

2.4.2 Expression of c-H-*ras* and c-K-*ras*

Both c-H-*ras* and c-K-*ras* sequences appear to be expressed ubiquitously. Nearly constant levels of c-H-*ras* and c-K-*ras* transcripts have been found during development of the mouse fetus (Fig. 1) and of extraembryonal structures (MÜLLER et al. 1982, 1983a, c) as well as in a variety of human tissues (MÜLLER et al. 1983b). Transcripts from c-H-*ras* (mouse) have been determined to be 1.4 kb (Fig. 6) (MÜLLER et al. 1982, 1983c). In some instances, low levels of 4.3-kb transcripts have also been observed. Transcripts related to c-K-*ras* (mouse) have been reported to be 2.0 kb and 4.4 kb (Fig. 6) (MÜLLER et al. 1983c). The sizes of c-H-*ras* and c-K-*ras* transcripts determined by ELLIS et al. (1982) in mouse cell lines are somewhat different, which is at least in part due to different sizes assumed for the 18S/28S rRNA markers. The level of v-*ras* transcripts is 10- to 30-fold higher in virus-transformed cells compared to the concentration of c-*ras*-related transcripts in uninfected cells (MÜLLER et al. 1982, 1983c).

Both c-H-*ras* and c-K-*ras* encode proteins of 21 K (P21) which have been detected at low levels in normal cells (relative to virus-transformed cells) from a variety of vertebrate species (LANGBEHEIM et al. 1980). A cell line derived from mouse bone marrow (416B), however, has been reported to contain markedly elevated concentrations (about 25-fold over the normal level) of c-K-*ras* transcripts (ELLIS et al. 1982) and to express abundant amounts of P21 (SCOLNICK et al. 1981). With regard to the physiological function of the c-K-*ras*-encoded P21, the significance of the high level of expression in 416B cells remains obscure, since elevated levels of c-K-*ras* transcripts or P21 have not been detected in normal bone marrow cells as yet. The relatively unspecific patterns of both c-H-*ras* and c-K-*ras* expression could suggest that their encoded proteins have general functions in normal cellular metabolism. A two- to four fold increase in the level of c-H-*ras* (rat) transcripts after partial hepatectomy concomitant with the burst of DNA synthesis has recently been reported (GOYETTE et al. 1983). This finding is consistent with the hypothesis that c-*ras* proteins may play a role in physiological growth processes.

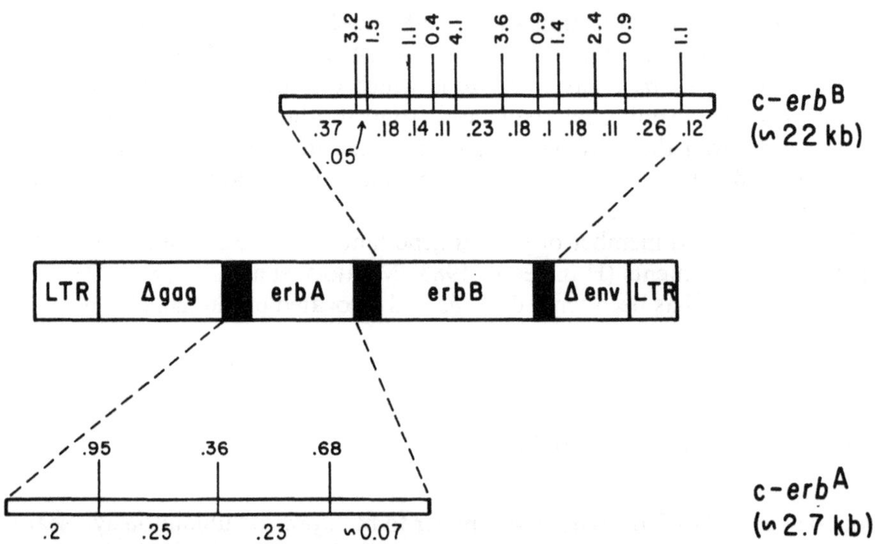

Fig. 11. Organization of erb^A and erb^B genes. The structure of AEV proviral DNA and the organisation of c-erb^A and c-erb^B in the chicken are shown. Introns are indicated by *vertical lines; open boxes* show the putative exons (VENNSTRÖM and BISHOP 1982). Numbers indicate kb of DNA

2.5 Two Oncogenes of Avian Erythroblastosis Virus: *erb*-A and *erb*-B

2.5.1 Characteristics and Expression of v-*erb* Genes

Avian erythroblastosis virus (AEV) causes erythroblastic leukemia and sarcomas in chickens, depending on the route of injection, and transforms both erythroid cells and fibroblasts in vitro (for review see GRAF and BEUG 1978). AEV contains two putative oncogenes, v-*erb*-A and v-*erb*-B (Fig. 11) (VENNSTRÖM and BISHOP 1982). The v-*erb*-A gene product has been identified as a 75 K gag-*erb*-A fusion protein (P75) that is translated from an AEV-genome-sized RNA (ANDERSON et al. 1980). In contrast, the v-*erb*-B gene is translated from a subgenomic RNA (ANDERSON et al. 1980; PRIVALSKY and BISHOP 1982) into a 62.5 K protein that is post-transcriptionally modified to yield a 68 K phosphorylated glycoprotein (HAYMAN et al. 1983). Analyses of v-*erb*-A and v-*erb*-B AEV deletion mutants revealed that the v-*erb*-B-encoded protein may be responsible for both induction of erythroblastosis in vivo and transformation of erythroid cells as well as fibroblasts in vitro (FRYKBERG et al. 1983). The results of these studies also suggested that the *erb*-A gene product may enhance the transforming potential of v-*erb*-B. The gag-*erb*-A gene product appears to be a cytoplasmic protein (BUNTE et al. 1982; ABRAMS et al. 1982), whereas the v-*erb*-B glycoprotein has been reported to be located in the cytoplasmic membrane of AEV-transformed erythroblasts with at least part of its exposed on the cell surface (HAYMAN et al. 1983). Chicken genomic sequences homologous to v-*erb*-A and v-*erb*-B have been shown to be independent cellular genes which are separated by at least 12 kb of DNA (VENNSTRÖM and BISHOP 1982). Two distantly related classes of c-*erb*-A genes have been identified in mouse and human DNA (JANSSON et al. 1983). The human c-*erb*-Al locus has been assigned to chromosome 17, whereas the human c-*erb*-B gene is on chromosome 7 (B. Vennström, personal communication).

2.5.2 Expression of c-*erb*

The c-*erb* genes have been reported to be transcribed in a variety of chicken cells and tissues with the highest concentration of transcripts observed in day-10 whole embryos (VENNSTRÖM and BISHOP 1982; GONDA et al. 1982a). Two transcripts from the c-*erb*-A gene of 3.0 kb and 4.5 kb have been identified in chicken embryos (VENNSTRÖM and BISHOP 1982). The c-*erb*-B mRNA species have been determined to be 9.0 kb and 12.0 kb, both of which are exceptionally large when compared with the v-*erb*-B coding region. It is thus possible that the c-*erb*-B mRNAs have uncommonly long 3′ untranslated regions. The c-*erb*-A and c-*erb*-B genes appear not to be expressed concordantly since c-*erb*-B transcripts are considerably less abundant in chick bone marrow, yolk sac, and thymus relative to c-*erb*-A-related RNA (quoted as T. Gonda, D. Sheiness, and B. Vennström, unpublished data, in Bishop and Varmus 1982).

In mice, expression of c-*erb*-A-specific sequences were detected at various stages of embryogenesis (days 7–18); a two- to fourfold enrichment of c-*erb*-A transcripts was observed in day-14–18 embryos as compared with day-7 embryos. c-*erb*-A-specific transcripts of 2.3 kb were identified in day-14 embryos. c-*erb*-B-specific transcripts were detected during mouse embryogenesis at very low levels (D. Slamon, personal communication).

2.6 The *myb* Gene of Avian Myeloblastosis Type Viruses

2.6.1 Characteristics and Expression of v-*myb*

Avian myeloblastosis virus (AMV) specifically induces myeloblastosis in chickens (Fig. 12) (for review see GRAF and BEUG 1978). Its oncogene has been termed v-*myb* (ROUSSEL et al. 1979). The oncogene product of AMV is translated

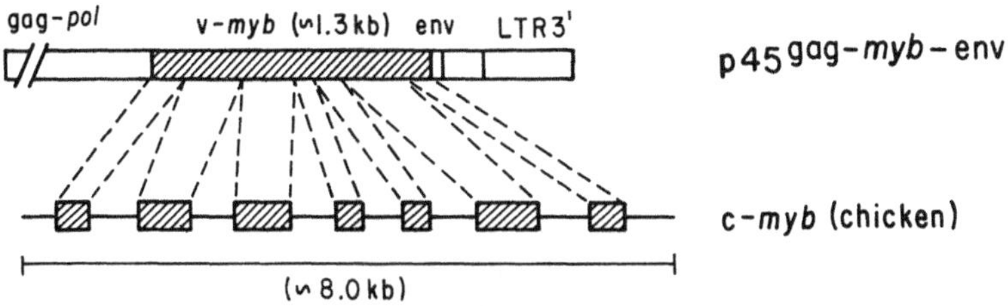

Fig. 12. Organization of *myb* genes. The structures of v-*myb* and its cellular homolog c-*myb* in chicken cells are shown. Localization of the exons/introns of the c-*myb* genome is based on nucleotide sequence analysis. (KLEMPNAUER et al. 1982)

from a subgenomic RNA that presumably contains only *myb*-specific sequences (CHEN et al. 1981; GONDA et al. 1981). Recently, a 48 K *myb*-specific protein was identified by using antisera against either synthetic peptides (BOYLE et al. 1983) or bacterial synthesized protein (KLEMPNAUER et al. 1983). Another AMV-type retrovirus, E26, has been reported to contain also the v-*myb* oncogene (ROUSSEL et al. 1979). Despite this similarity, E26 induces abnormal proliferation of both erythroid and myeloid precursor cells in vivo and in vitro (RADKE et al. 1982) and transforms fibroblasts in vitro (GRAF et al. 1979a). The putative transforming gene product of E26 is a gag-*myb*-x protein of 135 K (P135) (ROUSSEL et al. 1979; GRAF et al. 1979b). Like AEV, E26 seems to contain a second oncogene, referred to as *ets* (D. Stehelin, personal communication), which presumably encodes the unidentified portion, x, of the P135 protein. Cellular sequences (c-*myb*) homologous to v-*myb* have been identified (ROUSSEL et al. 1979; SOUZA et al. 1980; PERBAL and BALUDA 1982). Nucleotide sequence analyses indicate that the viral and cellular gene products differ at their C-termini (KLEMPNAUER et al. 1982). The v-*myb* protein has been localized in the cell nucleus (J.M. Bishop, personal communication), while the c-*myb* protein appears to be present in the cytoplasm (M. Baluda, personal communication).

Human c-*myb* sequences have been assigned to chromosome 6 (DALLA-FAVERA et al. 1982c; HARPER et al. 1983).

2.6.2 Expression of c-*myb*

In chickens, c-*myb* has been reported to be expressed at hight levels specifically in hematopoietic tissues, including yolk sac, bone marrow, thymus, and bursa (CHEN 1980; GONDA et al. 1982a). Relatively high levels of c-*myb* transcripts (comparable with the level in normal thymus) were also detected in AEV-transformed erythroblastic cells and in transformed cells from avian leukosis virus (ALV)-induced lymphomas (GONDA et al. 1982a). These results suggest that enhanced expression of c-*myb* is not restricted to cells of a single lineage, but probably occurs in immature cells of all hematopoietic differentiation pathways, i.e., in cells of lymphoid, myeloid, and erythroid origin (GONDA et al. 1982a). The decline in the level of c-*myb* transcripts observed in the bursa of older chickens could suggest a correlation of c-*myb* expression with proliferative activity of hematopoietic cells (CHEN 1980). GONDA et al. (1982a) stressed that there is no obvious correlation of the pattern of c-*myb* expression with the target cell/tissue specificity of AMV. Thus, a low level of c-*myb* expression was observed in macrophages, which represent target cells of AMV, and high levels of c-*myb* transcripts were detected in normal thymus cells, which are not susceptible to transformation by AMV. These data suggest that the determination of target cell specificity is a complex phenomenon that may not only depend on the availability of intracellular target proteins for the *myb* gene products.

Analysis of a variety of human leukemia and solid tumor cells (WESTIN et al. 1981a) showed results similar to those described for c-*myb* expression in chickens. c-*myb* transcripts were detectable only in leukemia cells which were similar to hematopoietic precursor cells of either the lymphoid, myeloid, or erythroid lineage. Furthermore, the level of c-*myb* transcripts in the promyelocytic cell line HL60 was found to decline to undetectable concentrations after induction of terminal differention by retinoic acid in vitro (WESTIN et al. 1982a). These results lend support to the hypothesis that c-*myb* gene expression and the physiological function of the c-*myb* protein may be correlated with the proliferation of hematopoietic precursor cells.

2.7 The *myc* Gene of Avian Myelocytomatosis Viruses

2.7.1 Characteristics of v-*myc*

Avian myelocytomatosis viruses induce a broad spectrum of malignancies in birds, including carcinomas, sarcomas, and, less frequently, myelocytomas. They also transform epitheloid, fibroblastic, and myeloid cells in vitro (for review see GRAF and BEUG 1978). Four independent isolates containing the v-*myc* oncogene have been described – MC29, MH2, CMII, and OK10 (Fig. 13; ROUSSEL et al. 1979). The MH2 virus contains an additional oncogene referred to as v-*mil* (JANSEN et al. 1983). The putative transforming gene products of these viruses are phosphorylated polyproteins that contain *onc*-sequences and part of the helper virus gag (MC29 and CMII) or gag and pol (OK10) sequences. Additionally, OK10 and MH2 viruses encode subgenomic RNAs containing only *myc* sequences. The MC29 gag-*myc* fusion proteins, p110, CMII fusion protein (p90), MH2 p61/63 *myc* protein, and OK10 p62 *myc* protein, have been reported to be located in the cell nucleus, as determined by using either gag antisera (ABRAMS et al. 1982; DONNER et al. 1982b; BUNTE et al. 1982) or, more recently, antisera against synthetic peptides (HANN et al. 1983). The MC29 gag-*myc* polyprotein has been reported to exhibit DNA binding activity in vitro (BUNTE et al. 1982). Recently, however, it has been shown that the *myc* protein is largely associated with the nuclear matrix (R. Eisenmann, personal communication). Cellular sequences homologous to v-*myc* have been identified in the genome of various vertebrate cells, including humans (SHEINESS and BISHOP 1979; SHEINESS et al. 1980; EVA et al. 1982). Using antisera raised against synthetic peptides, the c-*myc* protein has been shown to be 60 K (HANN et al. 1983). Like the v-*myc* polyproteins, it has also been located in the cell nucleus (HANN et al. 1983). Data obtained from sequence analysis of c-*myc* cDNA clones from human cells predict a size of 48 K (COLBY et al. 1983; WATT et al. 1983). Presumably, the c-*myc* product, which is rich in proline residues, shows anomalous (slower) migration on SDS-PAG.

The human c-*myc* gene has been assigned to chromosome 8 (DALLA FAVERA et al. 1982; NEEL et al. 1982; TAUB et al. 1982), whereas the mouse homolog is located on chromosome 15 (CREWS et al. 1982).

2.7.2 Expression of c-*myc*

The c-*myc* gene appears to be expressed in a great variety of cells and tissues in birds (SHEINESS et al. 1980; GONDA et al. 1982a), mice (D.J. Slamon and

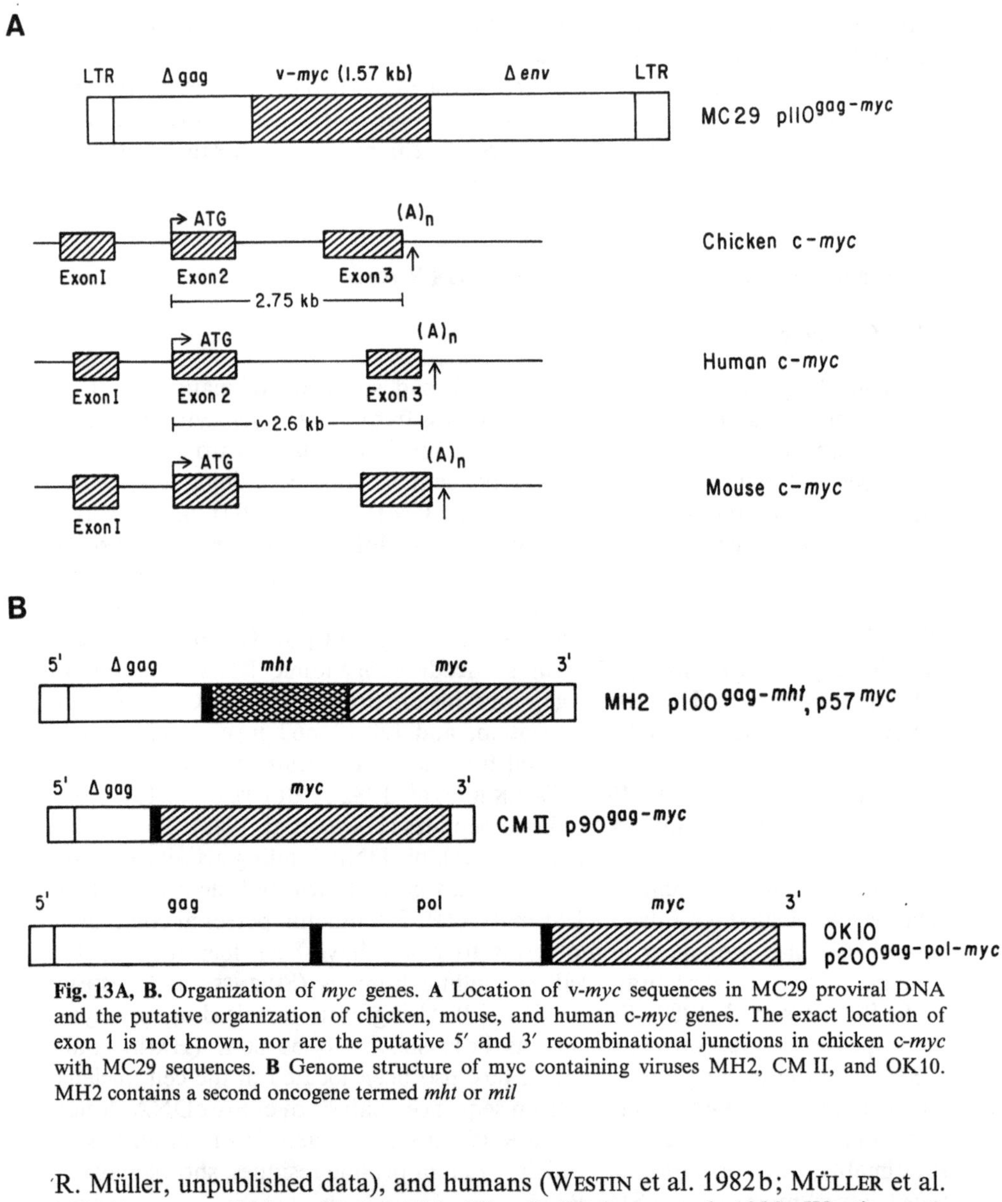

Fig. 13A, B. Organization of *myc* genes. **A** Location of v-*myc* sequences in MC29 proviral DNA and the putative organization of chicken, mouse, and human c-*myc* genes. The exact location of exon 1 is not known, nor are the putative 5′ and 3′ recombinational junctions in chicken c-*myc* with MC29 sequences. **B** Genome structure of myc containing viruses MH2, CM II, and OK10. MH2 contains a second oncogene termed *mht* or *mil*

R. Müller, unpublished data), and humans (Westin et al. 1982b; Müller et al. 1983b) as an mRNA of 2.5–2.7 kb (Fig. 4) (Sheiness et al. 1980; Westin et al. 1982b; D. Slamon and R. Müller, unpublished data). However, the level of c-*myc* transcripts in chicken hematopoietic tissues has been reported to be elevated (approximately 10- to 30-fold) compared with other tissues, such as brain, liver, and muscle (Gonda et al. 1982a). Readily detectable levels of c-*myc* transcripts have also been described in a variety of human leukemia cell lines which are similar to immature hematopoietic cells (Westin et al. 1982b). These observations could suggest that the c-*myc* gene product plays a role in normal hematopoiesis, although its physiological function may not be restricted to hematopoietic cells, since c-*myc* expression has also been detected in other chicken, mouse, and human tissues (GONDA et al. 1982a; MÜLLER et al. 1983b). Elevated levels of c-*myc* transcripts have also been reported in some mouse plasmacytomas and human Burkitt lymphoma cells (see section 3).

During mouse prenatal development, c-*myc*-related transcripts were detected at all stages from days 7 to 18, but higher concentrations were detected at day 17 and 18. Relatively high levels of c-*myc* expression appear to occur specifically in embryonic tissue, since few or no c-*myc*-related transcripts were detected in day-18 placental tissue. A single c-*myc* specific transcript of 2.7 kb could be detected in day 17 mouse embryos (D. Slamon, personal communication).

2.8 The Oncogenes of Avian Sarcoma Viruses

2.8.1 Characteristics and Expression of v-*src*, v-*fps*, v-*yes*, and v-*ros*

Avian sarcoma viruses (ASV) induce predominantly fibrosarcomas as well as a variety of other neoplasms in chickens and mammals, depending on the site of inoculation, and transform fibroblasts in vitro (for review see TEICH et al. 1982; BISHOP and VARMUS 1982). Four different oncogenes of ASVs have been identified and isolated to date: v-*src* (BRUGGE and ERIKSON 1977; PURCHIO et al. 1978; BRUGGE et al. 1979; of Rous sarcoma virus (RSV); v-*fps* (SHIBUYA et al. 1980; WANG et al. 1981) of Fujinami sarcoma virus (FSV), URI-ASV, PRCII-ASV, and PCRIV-ASV; v-*yes* (KAWAI 1980; YOSHIDA et al. 1980, 1981) of Y73-ASV and Esh sarcoma virus (ESV); and v-*ros* (WANG et al. 1982; SHIBUYA et al. 1982) of UR2-ASV (Fig. 14). The v-*fps* gene has been shown to be homologous to v-*fes*, the oncogene of GA/ST FeSV (see section 2.4.) (BARBACID et al. 1981; SHIBUYA et al. 1980, 1982; HAMPE et al. 1982). The gene product of v-*src* is a 60 K phosphoprotein (p60) that does not contain helper virus-derived sequences. In contrast, the transforming gene products of v-*fps*, v-*yes*, and v-*ros* are all gag-*onc* fusion proteins of 68–170 K (for review see BISHOP and VARMUS 1982). All these transforming proteins are associated with tyrosine protein kinase activity (COLLETT and ERIKSON 1978; HUNTER and SEFTON 1980; for review see COOPER and HUNTER 1983). The bulk of the v-*src* protein p60 in RSV-infected cells has been shown to be associated with the plasma membrane, where it appears to be in close affiliation with the cytoplasmic aspect, particularly in areas of tight junctions with adjacent cells (WILLINGHAM et al. 1979; ROHRSCHNEIDER 1979, 1980; COURTNEIDGE et al. 1980; LEVINSON et al. 1981). Several putative cellular targets of the p60 *src* Kinase activity have been identified (COOPER and HUNTER 1983; SEFTON et al. 1981b). Like two other plasma-membrane-bound v-*onc* proteins (see sections 2.3 and 2.4), p60 has been reported to be modified with lipid which may be one of the parameters determining its intracellular location (SEFTON et al. 1982; GARBER et al. 1983). The complete nucleotide sequence of Y73 viral DNA has been determined (KITAMURA et al. 1982), and a remarkable homology between *yes* and *src* genes was uncovered. Not surprisingly, the intracellular distribution of the $p90^{gag\text{-}yes}$ or $p80^{gag\text{-}yes}$ (ESV) proteins appears to be similar to that of

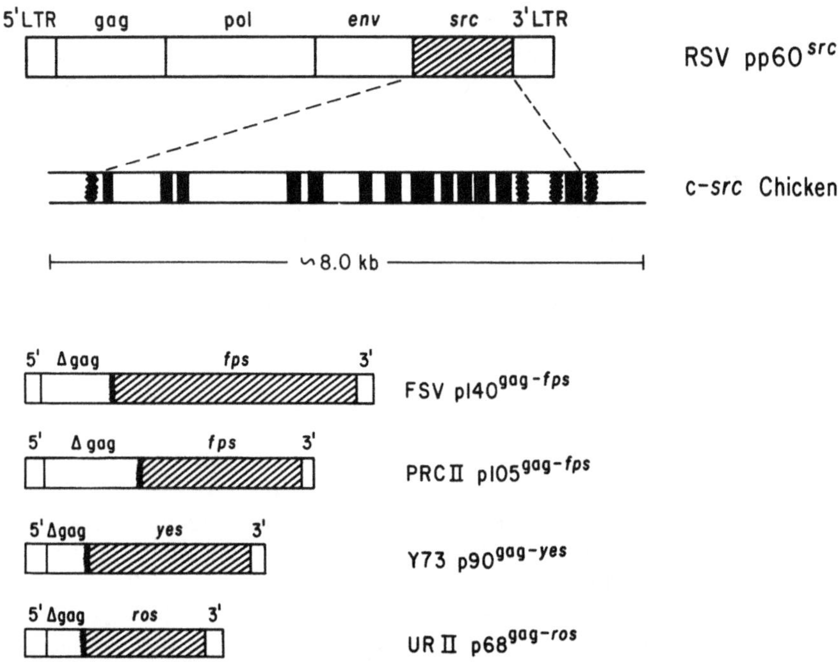

Fig. 14. Organization of oncogenes of avian sarcoma viruses. The structure of chicken c-*src* based upon heteroduplex analysis (PARKER et al. 1981; TAKEYA et al. 1981) and nucleotide sequence analysis (TAKEYA and HANAFUSA 1983) is shown. The exact size of the exons (*black boxes*) is not known, nor is the recombinational junction with RSV. The total size of the c-*soc* coding domain is indicated (~8.0kb). The putative structures of Fujinami sarcoma virus (*fps*/*fes*). PRC II (*fps*/*fes*), Y-73 (*yes*), and URII (*ros*) are shown. Organization of the cellular homolog of *fps* is similar to that of *fes* (see Fig. 7)

pp60^{src} (ROHRSCHNEIDER and GENTRY 1983). Quantitative cellular fractionation of FSV-transformed cells revealed that p140$^{gag\text{-}fps}$ is associated with plasma membrane and cytoplasmic components (FELDMAN et al. 1983). No subcellular localization for the p68$^{gag\text{-}ros}$ protein has yet been reported.

2.8.2 Expression of c-*src*, c-*fps*, c-*yes*, and c-*ros*

The chicken genomic homolog, c-*src*, of v-*src* was the first cellular counterpart of a retroviral oncogene to be identified (STEHELIN et al. 1976), as well as the first c-*onc* gene found to be expressed in normal cells (SPECTOR et al. 1978a, b). Expression of c-*src* has been reported not to be correlated with proliferative activity of chicken fibroblastic cells in vitro (SPECTOR et al. 1978b). Analyses of a variety of chicken tissues have shown that c-*src* is expressed ubiquitously (GONDA et al. 1982a; SHIBUYA et al. 1982), with the highest levels of c-*src* transcripts found in macrophages, lung, and spleen, followed by other hematopoietic tissues, fibroblasts, and whole chicken embryos (GONDA et al. 1982a; SHIBUYA et al. 1982). The level of c-*src* transcripts in spleen has been determined to be 1.4 copies per cell (SHIBUYA et al. 1982) or 0.006% of the total cellular polyadenylated RNA (GONDA et al. 1982a). High levels of c-*src* protein have been detected in neural tissues (BARNEKOW et al. 1982; COTTON and BRUGGE 1983). Expression of c-*src* protein has been observed in retinal cells during embryonic chick development (P. MANESS, personal communication). These findings may suggest a function of the c-*src* gene product in cells of neuroectodermal origin. The chicken c-*src* gene is expressed as a 3.9-kb mRNA (GONDA et al. 1982), which is translated into a 60 K phosphoprotein associated with tyrosine protein kinase activity (COLLETT et al. 1978, 1979; BRUGGE et al. 1979; OPPERMANN et al. 1979). c-*src*-specific transcripts were observed at all stages of mouse embryogenesis, though higher concentrations of transcripts could be detected during the latter half of embryonic development. Size analysis of poly(A)$^+$ RNA from the late stages of embryos identified a c-*src* transcript of 4.0 kb (D. Slamon, personal communication).

Transcripts from the c-*yes*, c-*fps*, and c-*ros* genes have been detected in most chicken tissues analyzed (SHIBUYA et al. 1982). In general, c-*yes* transcripts were observed to be more abundant than c-*fps*- and c-*ros*-related RNA. The highest level of c-*yes* expression was observed in normal kidney (26 RNA copies per cell). The concentration of c-*yes* transcripts was 5- to 26-fold greater in kidney than in any other tissue examined. The highest level of c-*fps* transcripts was detected in bone marrow (2.5 transcripts per cell), whereas the highest concentration of c-*ros*-related RNA was observed in kidney (2.5 copies per cell). In normal chicken bone marrow cells, a 98 K protein could be identified that is immunologically cross-reactive with the FSV P140 protein (MATHEY-PREVOT et al. 1982). Currently available data on expression of c-*yes*, c-*fos*, and c-*ros* are, however, inadequate to allow speculation as to the functions of the cellular homologs of ASV oncogenes.

The human homolog of v-*fps*/v-*fes* has been assigned to chromosome 15 (DALLA FAVERA et al. 1982c; HEISTERKAMP et al. 1982; HARPER et al. 1983), while the human c-*src* gene has been localized on chromosome 20 (for reference, see Barker et al. 1983).

2.9 The *rel* Gene of Avian Reticuloendotheliosis Virus

The avian reticuloendotheliosis virus strain T (REV-T) causes rapid leukemias in hatched chicks and turkey poults (THEILEN et al. 1966). The REV-T oncogene v-*rel* has a counterpart in the normal vertebrate genome (WONG and LAI 1981; CHEN et al. 1982). The c-*rel* sequences are expressed in normal chicken spleen as a 4.0-kb mRNA (CHEN et al. 1983). No analyses of stage- or tissue-specific expression of c-*rel* have been published to date.

2.10 The c-*mos* Locus

The Moloney murine sarcoma virus (Mo-MSV) (MOLONEY 1966) induces rhabdomyosarcomas and transforms fibroblasts in vitro. The transforming gene v-*mos*

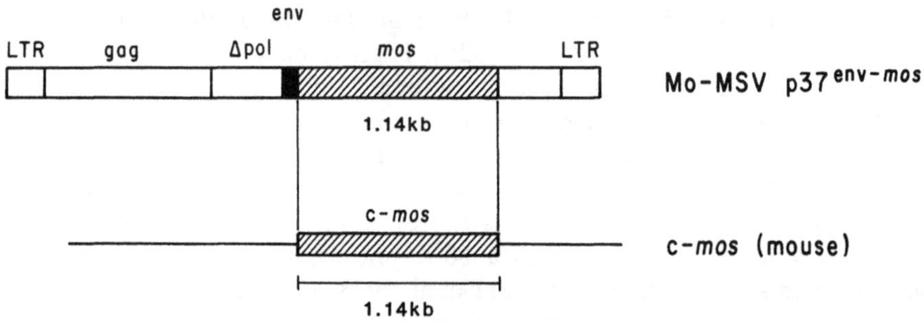

Fig. 15. Organization of *mos* genes. Structures are based on nucleotide sequence analyses of Mo-MSV and c-*mos* (mouse) (VAN BEVEREN et al. 1981)

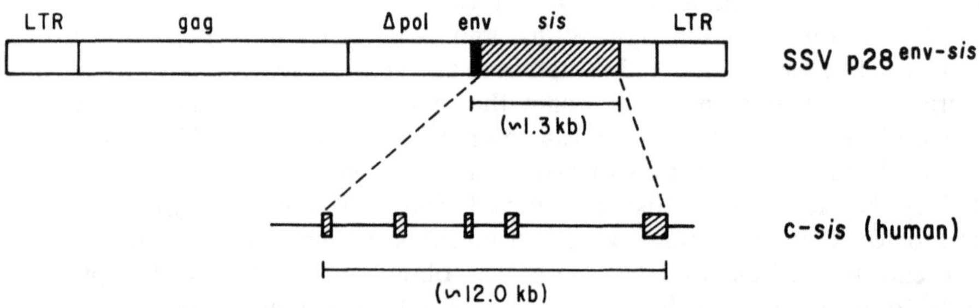

Fig. 16. Organization of *sis* genes. The structure of simian sarcoma virus proviral DNA (DEVARE et al. 1983) and the organization of c-*sis*-specific sequences in human cells (DALLA-FAVERA et al. 1981) are shown

of Mo-MSV and its cellular homolog c-*mos* have been extensively characterized (Fig. 15) (VAN BEVEREN et al. 1981). Using antibodies raised against the predicted C-terminal amino acids of the v-*mos* protein, a 37 K cytoplasmic protein has been identified in Mo-MSV-transformed cells (PAPKOFF et al. 1981). Despite a sustained effort by a number of laboratories, no c-*mos*-specific transcripts have been identified in a variety of normal tissues and cell lines (FRANKEL and FISCHINGER 1976; GATTONI et al. 1981; MÜLLER et al. 1982; MÜLLER 1983). It has been proposed that c-*mos* sequences are hypermethylated, which may prevent their transcription (GATTONI et al. 1981). It is also possible that c-*mos* sequences do not represent a complete transcriptional unit, since no transcription initiation and termination signals could be detected by nucleotide sequence analysis (VAN BEVEREN et al. 1981). Recently, mouse c-*mos* sequences have been reported to be expressed in neoplastic cells as the result of a genomic rearrangement (RECHAVI et al. 1982) (see Sect. 3). Another mouse sarcoma virus, the Gazdar MSV (GAZDAR et al. 1972), has been shown to contain also an oncogene derived from c-*mos* sequences.

The human homolog of v-*mos* has been assigned to chromosome 8 (PRAKASH et al. 1982; NEEL et al. 1982).

2.11 The *sis* Gene of Simian Sarcoma Virus

The simian sarcoma virus (SSV) was isolated from the fibrosarcoma of a wooly monkey. The structure of the SSV genome and the cellular homolog of its transforming gene *sis* are shown in Fig. 16 (DALLA-FAVERA et al. 1981; DEVARE et al. 1983). Recently, a 28 K env-*sis* gene product was identified in SSV-transformed cells (ROBBINS et al. 1983; DEVARE et al. 1983). It appears that $p28^{sis}$ is glycosylated, based upon the presence of tunicamycin-sensitive sites (K. ROBBINS and S. AARONSON, personal communication). No c-*sis*-specific transcripts have thus far been found in normal cells or tissues. However, expression of c-*sis* could be detected in a variety of human tumor cell lines, where transcripts of 4.2 kb have been identified (EVA et al. 1982; WESTIN et al. 1982b).

One of the fascinating findings of oncogene research is the recent discovery of an extensive homology between the v-*sis* and platelet-derived growth factor (PDGF) amino acid sequences (WATERFIELD et al. 1983; DOOLITTLE et al. 1983). This observation suggests that c-*sis* is the gene encoding PDGF. Although growth factor-inducing properties have been detected with several other v-*onc* genes (DELARCO and TODARO 1978; TWARDZIK et al. 1982; quoted in WEISS 1983), this finding represents the first direct evidence for the function of a c-*onc* gene and may elucidate the mechanism of neoplastic transformation induced by its viral counterpart.

The human homolog of v-*sis* has been assigned to chromosome 22 (DALLA-FAVERA et al. 1982d).

3 Amplification, Rearrangement, and Translocation of c-*onc* Sequences in Tumor Cells

Specific karyotypic changes often accompany human neoplasia (KLEIN 1981, 1983). Chromosomal aberrations have been repeatedly observed in Burkitt's lymphoma, salivary gland tumors, small-cell carcinomas of the lung, meningiomas, acute non-lymphoblastic leukemia, and other hematopoietic malignancies (MITELMAN and LEVAN 1981). Some of these karyotypic changes are associated with two types of alterations – c-*onc* gene amplifications and translocations.

3.1 Gene Amplification and Rearrangements

In the human promyelocytic cell line HL60, a tenfold amplification of the human c-*myc* gene has been reported (COLLINS and GROUDINE 1982; DALLA-FAVERA et al. 1982a). The c-*myc* amplification appears to be localized to a homogenous staining region (HSR) of chromosome 8, which has previously been shown to be the normal site of the c-*myc* locus (NEEL et al. 1982; DALLA-FAVERA et al. 1982b; R.C. Gallo, personal communication). The c-*myc* amplification in HL60 cells is accompanied by a significant increase in the level of c-*myc* transcripts (WESTIN et al. 1982b). Since the same c-*myc* amplification was detected in fresh leukemic cells of the same individual from whom the HL60 cell line orginated, DALLA-FAVERA et al. (1982a) suggested that c-*myc* amplification may have been involved in the induction of leukemia in this case. This hypothesis will, however, be difficult to prove since nonhematopoietic cells from the same patient are not available for analysis. It can, thus, not be ruled out that c-*myc* is also amplified in germ cells and hence in all somatic cells (such as the c-H-*ras* amplification in *Mus pahari:* see section 2.4.1) without any adverse effects on normal cellular metabolism. Alternatively, c-*myc* amplification might represent a normal event in the differentiation of myelocytic cells (DALLA-FAVERA et al. 1982a). Amplification of the c-*myc* gene was also observed in cell lines derived from human colon carcinoma, where the amplified c-*myc* sequences were found on double minute chromosomes or within homogenously staining regions on a distorted X chromosome (ALITALO et al. 1983).

A 30- to 60-fold amplification of the c-K-*ras* gene was found in a mouse adrenocortical tumor (SCHWAB et al. 1983). The concentrations of c-K-*ras*-encoded transcripts and of p21 were shown to be similarly elevated. The amplified c-*onc* sequences could be assigned to double minute chromosomes and homogenously staining chromosomal regions, which are common karyotypic characteristics of tumor cells (SCHWAB et al. 1983).

A four- to eight-fold amplification of human c-*abl* sequences has been observed in the human chronic myelogenous leukemia cell line K562 (COLLINS and GROUDINE 1983). In addition, restriction analysis of K562 DNA revealed a rearrangement of c-*abl* sequences in this cell line. Moreover, relatively high levels of c-*abl* transcripts were observed in K562 cells compared with other normal and leukemic cells. Since the K562 cell line contains the Philadelphia chromosome (t9:22), it is noteworthy that human c-*abl* is located on chromosome 9 and human immunoglobulin (Ig) lambda light-chain control region has been assigned to chromosome 22. Interestingly, the Ig lambda light-chain genes in K562 are also amplified but apparently are not rearranged and show no enhanced expression. Rearrangements of the c-*myb* locus has also been detected in A-MuLV induced plasmacytoid lymphosarcomas (MUSHINSKI et al. 1983). These tumors do not contain any detectable A-MuLV proviral DNA but express abundant c-*myb* RNA of unusually large size.

3.2 Translocations

A substantial body of evidence indicates that chromosomal translocations involving the c-*myc* oncogene and one of the Ig loci are quite frequent in B-cell tumors of mouse or human origin (KLEIN 1983; PERRY 1983). Studies of both Burkitt's lymphoma and mouse plasmacytomas have shown that translocation breakpoints in the chromosome carrying the c-*myc* gene (chromosome 15 in mice, chromosome 8 in humans) and those carrying the Ig genes can vary among different tumors. In over 80% of the mouse plasmacytomas at least one *myc* allele has undergone a recombination within the 11-kb *Eco* RI fragment containing the c-*myc* gene (HARRIS et al. 1982; SHEN-ONG et al. 1982; ADAMS et al. 1982). The breakpoints generally occur within a 1.0-kb region that spans exon 1 of the c-*myc* gene and the intron separating it from exon 2 (ADAMS et al. 1983; STANTON et al. 1983) (Fig. 17). In Burkitt's lymphoma, heterogeneity of the breakpoint is also observed since some breakpoints occur within or near exon I whereas others occur 8 kb or more upstream (TAUB et al. 1982; ADAMS et al. 1983). A common recombination locus appears to be the switch region of Ig heavy chain (chromosome 14 in humans and 12 in the mouse). In a lower proportion of cases, translocation of c-*myc* to chromosome 2 (kappa light chain) or chromosome 22 (lambda light chain) has also been found (TAUB et al. 1982; ERIKSON et al. 1983).

c-*myc* translocations result in two types of change: (a) elevated levels of c-*myc*-related transcripts, and (b) aberrant c-*myc* transcripts in mouse plasmacytomas. Levels of c-*myc* expression are usually increased in mouse plasmacytomas and Burkitt's lymphomas, but this transcriptional activation varies from two- to tenfold. It may be worth noting that in contrast, nearly 20- to 30-fold higher levels of c-*myc* transcripts are observed in ALV-induced B-cell lymphomas, compared with normal hematopoietic tissue (HAYWARD et al. 1981; GONDA et al. 1982a). The nature of the c-*myc* protein in tumors containing translocations has not been studied, but in mouse plasmacytomas an altered form of the c-*myc* transcript, which is 0.4 kb shorter than the normal c-*myc* RNA, was detected (SHEN-ANG et al. 1982). The truncated mRNA contains c-*myc* exons 2 and 3 and a segment derived from the first intron. Since the first exon appears to contain terminator codons in all reading frames, the translational initiation codon was inferred to be at the beginning of exon 2 (STANTON et al. 1983). Thus the truncated and nontruncated *myc* mRNAs would be expected to encode the same *myc* protein. STANTON et al. (1983) have suggested that shortening the 5′ end of *myc* RNA, which contains an unusually long 5′ untranslated region (over 0.4 kb), might result in greater translation efficiency and hence larger amounts of c-*myc* protein.

Identification of transcriptional enhancer elements in immunoglobulin heavy-chain and light-chain genes suggests that enhanced transcription of the rearranged c-*myc* gene might be due to these specific enhancers (BANERJE et al. 1983; GILLIES et al. 1983; QUEEN and BALTIMORE 1983; NEUBERGER 1983). However, the localization of the enhancer sequences in regions upstream of the Ig switch region (S) poses an obstacle, since this region will not necessarily be juxtaposed next to c-*myc* sequences after 5′-5′ recombinations between *myc* and any of the switch sequences (Fig. 17). Only in translocations where recombination occurs several kilobases upstream of the c-*myc* promoter, could enhancer elements still be functional.

In human chronic myelocytic leukemia (CML), c-*abl* sequences (see section 2.3) have been reported to be translocated from their normal position on chromosome 9 to the Philadelphia chromosome $22q^-$ by a direct reciprocal exchange between the two chromosomes (DE KLEIN et al. 1982). The significance of this finding is at present difficult to assess, since it remains unknown whether the level of c-*abl* expression or the structure of the c-*abl* gene gene product(s) is altered in CML cells.

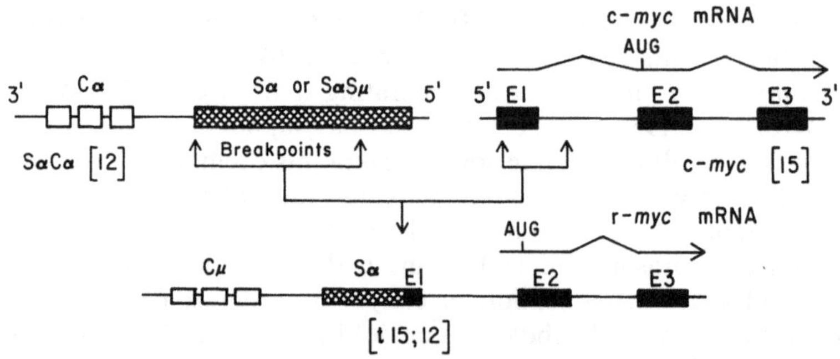

Fig. 17. Proposed structure of normal and rearranged mouse *myc* genes. The figure displays the organization of the normal c-*myc* gene on chromosome 15 (top right) and of the rearranged r-*myc* gene (bottom) after translocation into the immunoglobulin locus SαCα on chromosome 12 (top left). Exons (E1, E2, E3) are shown as *black boxes*. Numbers in parentheses are the chromosomal locations of c-*myc*, SαCα and r-*myc*. The possible structures of normal and rearranged c-*myc* mRNA are indicated. (Adapted from Perry 1983)

4 Mechanisms of Neoplastic Transformation: Role of c-*onc* Gene Products

The broad spectrum of neoplasias affecting a wide variety of cell types may involve multiple mechanisms of transformation. Three widely held views on possible mechanisms of neoplastic conversion are discussed in the following sections.

4.1 Quantitative Modulations

When the product of a viral or cellular oncogene is synthesized at elevated levels, this may affect the normal cellular metabolic processes in a manner that leads to cellular transformation. One of the first examples of this mode of oncogenesis was the finding that c-*mos* (mouse) sequences could transform NIH/3T3 cells when linked to a viral long terminal repeat (LTR) (Oskarsson et al. 1981; Blair et al. 1981). Similarly, addition of LTR sequences to the c-H-*ras*-1 (human) gene can lead to cellular transformation (Chang et al. 1982a). The LTR contains both a promoter and an enhancer element (Jolly et al. 1983). Recently, G. Vande Woude and colleagues have shown that the linking of enhancer elements from the MSV-LTR to *mos* sequences is sufficient to induce transformation of fibroblasts (G. Vande Woude, personal communication). In mouse myeloma cells, Rechavi et al. (1982) have shown that c-*mos* sequences are rearranged such that their 5′ sequences are substituted by a 159-bp DNA segment corresponding to the LTR of an intracisternal A-type particle. As a consequence of this transposition, the myeloma cells express 1.4-kb mRNA containing c-*mos* sequences, which are usually transcriptionally inactive (Section 2.10).

Avian leukosis virus (ALV) induces B-cell lymphomas which is thought to result at least in part from activation of the cellular *myc* gene (HAYWARD et al. 1981; PAYNE et al. 1982). Over 80% of all ALV-induced B-cell neoplasias were found to contain ALV-LTR sequences integrated in the vicinity of the c-*myc* gene (HAYWARD et al. 1981). Integration of a provirus next to the c-*myc* gene has also been observed in 90% of B-cell lymphomas induced by nondefective reticuloendotheliosis virus (NOORI-DALOII et al. 1981). In the case of B-cell lymphomas, the LTR sequences have been shown to be present not only upstream but also downstream from the c-*myc* gene (PAYNE et al. 1982). HANN et al. (1983) described a c-*myc*-related 62 K phosphoprotein in bursal lymphoma cells, while in normal B cells only a nonphosphorylated form of a 60 K c-*myc* protein could be observed. Recently, it was reported that in ALV-induced erythroblastosis in chickens, the ALV provirus had integrated next to another c-*onc* gene, namely c-*erb*-B (FUNG et al. 1983). The available data thus strongly suggest that transcriptional activation of c-*myc* and c-*erb*-B genes are implicated in the induction by ALV of B-cell lymphomas and erythroblastosis, respectively.

When DNA from ALV-induced lymphomas were transfected onto NIH/3T3 cells, a transforming gene other than c-*myc* was identified (COOPER and NEIMAN 1981; GOUBIN et al. 1983). Molecular analysis of this transforming gene, designated *B-lym,* indicates a maximum coding capacity of 65 amino acids. Antibodies raised against synthetic B-*lym*-specific peptides recognize a protein of approximately 8–10 K in bursal lymphoma cells (G. COOPER, personal communication). It is possible that the activated c-*myc* gene and the B-*lym* gene act at different stages of the multistage process of neoplastic transformation of B-cells. It is interesting that, like the c-*myc* gene product, the B-*lym* protein is also located in the cell nucleus (G. COOPER, personal communication).

4.2 Qualitative Alterations

When an oncogene product undergoes a mutation with respect to the normal cellular protein this may lead to the activation of its oncogenic potential. A number of transforming genes have been isolated from mouse and human tumor cells by transfection assays using NIH/3T3 recipient cells (for review see COOPER 1982). Many of them appear to be homologs to previously identified v-*onc* genes. For instance, the transforming genes isolated from human bladder and lung carcinoma cell lines are homologous to the *ras* genes of Harvey and Kirsten sarcoma viruses (C-H-*ras*-1 and C-K-*ras*-1) (DER et al. 1982; PARADA et al. 1982; SANTOS et al. 1982). Molecular analysis of the oncogene of the human bladder carcinoma cell line EJ or T24 revealed that a single amino acid substitution near the N-terminus (position 12) is responsible for the activation of its transforming potential (TABIN et al. 1982; REDDY et al. 1982; TAPAROWSKY et al. 1982). As a consequence of this point mutation, the c-H-*ras*-encoded P21 showed a slightly altered electrophoretic mobility, but no enhanced level of expression (TABIN et al. 1982). Similar findings have been reported for the P21 in human colon and lung tumor cells containing activated c-K-*ras* genes (DER and COOPER

1983; SHIMIZU et al. 1983a). Recently, a human lung carcinoma cell line was shown to contain an activated c-H-*ras* oncogene with a single amino acid substitution at position 61 (YUASA et al. 1983). It thus appears that structural alterations of c-*onc* proteins represent a way of activating their oncogenic potential. The activated H-*ras* oncogene of EJ cells, however, apparently is not sufficient to induce neoplastic transformation of primary cells. In recent studies it was demonstrated that malignant cells could be generated only when the activated H-*ras* gene was cotransfected with a second gene, such as *myc*, polyoma virus large T antigen (LAND et al. 1983) or adenovirus E1a sequences (RULEY 1983). These findings strengthen the view of malignant transformation as a multi-stage process (for review, see FARBER and CAMERON 1980). A third member of the *ras* family, N-*ras*, has been reported to be activated in human sarcoma, leukemia, and neuroblastoma cell lines (HALL et al. 1983).

In several instances it has been shown that v-*onc* genes and their cellular homologs have structurally different C-termini. For example, nucleotide sequence analyses of v-*myb* and c-*myb* genes have indicated that the predicted *onc* gene products differ at their C-termini (KLEMPNAUER et al. 1982). Similarly, the major difference between the deduced amino acid sequences of $pp60^{c\text{-}src}$ and $pp60^{v\text{-}src}$ is that 19 C-terminal amino acids of $pp60^{c\text{-}src}$ are replaced by a new sequence of 12 amino acids in $pp60^{v\text{-}src}$. If the C-terminus of v-*src* is switched with the C-terminus of c-*src*, no transformation activity can be detected, even though the associated protein kinase activity is evident (D. SHALLOWAY, personal communication). In contrast, the v-*fos* and c-*fos* proteins differ at their C-termini, but these differences do not appear to influence their transforming potential (MILLER et al. 1984). In cases where the *onc* genes form a fusion product (for instance, *gag-myc, gag-abl, env-sis, env-myb*), the corresponding cellular protein would be different in size and structure.

4.3 Cell Type Specificity

Expression of a cellular or viral oncogene at abnormal levels or in a structurally aberrant form may not be sufficient to induce malignant transformation. In addition, the induction of neoplastia may require expression of the oncogene in a cell type that is susceptible to malignant conversion by the respective oncogene product. Such target cell requirements appear to apply to many leukemia virus oncogenes (such as *myb*; see section 2.6), and have been well documented in the case of *fos* (see also section 2.1). The level of c-*fos* transcripts in the late-gestation amnion is nearly equal to the concentration of v-*fos* mRNA in FBJ-MSV-transformed cells (MÜLLER et al. 1983a, b). c-*fos* protein can be detected in day 18 mouse amnion by immunofluorescence (CURRAN et al. 1984). However, such a high level of c-*fos* expression in the amnion does not lead to malignant transformation although the c-*fos* gene product can induce neoplastic transformation of other cell types (MILLER et al. 1984). These results, therefore, strongly suggest that only certain cell types represent target cells for the *fos* protein and most probably for other c-*onc* gene products. Such target cells could be those cell types where the respective c-*onc* gene has a physiological function. These cells could provide intracellular target molecules with which the *onc* protein needs to interact. If the *onc* gene product is expressed constitutively at abnormal levels or has an aberrant (mutated) structure, it may escape the cellular regulatory mechanism and, as a consequence, initiate the process of neoplastic transformation. It is also possible that those cells that represent targets for *onc* gene products do not physiologically express the normal c-*onc* protein. Provided that suitable intracellular target molecules are available, such cells could be transformed as a consequence of the lack of regulatory elements for the respective oncogene product. These hypothetical models clearly demonstrate that the identification of those cell types in which c-*onc* genes are expressed and have physiological functions as well as of those cells that are susceptible to neoplastic transformation by oncogene products ("target cells") remain challenging tasks of present day oncogene research. It should also be noted that the issue is further complicated by the fact that neoplastic transformation is generally considered to be a multi-stage process and may require the activation of additional (onco)genes at different stages.

5 Summary and Prospects

The oncogenes of acutely transforming retroviruses have been acquired from host chromosomal DNA. The role of these sequences in cellular transformation and the function of their cellular homologs (c-*onc*) in normal cellular metabolic processes have been a subject of intense interest but remain poorly understood. Using molecularly cloned *onc*-specific probes, tissue-, cell-type- and stage-specific expression of a number of c-*onc* genes have been studied. Table 2 summarizes the extent of our current knowledge regarding the tissue specificity of c-*onc* gene expression, the nature of the c-*onc* transcripts, and the size of their products. Two main themes emerge: (a) During prenatal development, several c-*onc* genes are expressed in a stage- and tissue-specific manner. For instance, c-*fos* and c-*fms* are predominantly expressed in ancillary embryonic tissues, whereas c-*abl* is expressed at relatively high levels in the fetus. In contrast, the c-*ras* genes are expressed at all stages investigated in both embryonal and extra-embryonal tissues. (b) In some instances, tissue-specific expression could be detected during post-fetal development. Thus, c-*fos* is expressed at considerably higher levels in neonatal "bone" when compared to other tissues such as brain, liver, spleen, or bone marrow cells. c-*abl* expression can be detected at high levels specifically in the testes and in some hematopoietic tissues. Present data, however, do not allow definitive correlations between the tissue expressing a particular c-*onc* gene and the type of tumor induced by the corresponding v-*onc* gene.

Differential expression of c-*onc* genes combined with their extensive phylogenetic conservation lends credence to the widely held notion that their products exert functions in normal cellular metabolism. The present data, however, cannot unequivocally assign a certain physiological function to any of the c-*onc* genes identified. Two types of experimental approaches are needed: (a) Identifi-

Table 2. Properties of c-*onc* genes[a]

c-onc gene	Chromosomal location		Transcripts		Putative gene product (K)[b]
	Mouse	Human	Size (kb)	Tissue with highest expression (species)	
c-*src*	?	20	3.9–4.0	Spleen (ck), macrophages (ck), embryo (m)	60
c-*fps*/c-*fes*	7	15	?	Bone marrow (ck)	98/92
c-*ros*	?	?	?	Kidney (ck)	?
c-*yes*	?	?	?	Kidney (ck)	?
c-*myb*	?	6	4.0–4.5	Thymus (ck), myeloblasts (h)	
c-*myc*	15	8	2.5–2.7	Spleen, thumus, bursa (ck), fetus (m)	60 (48)[c]
c-*erb*-A	?	17	3.0+4.5	Embryo (ck)	?
c-*erb*-B	?	7	9.0+12.0	Embryo (ck)	?
c-*rel*	?		4.0	[Spleen] (ck)	?
c-*abl*	2	9	4.7+5.7 3.7+4.7+5.7	Thymus (m) Testes (m)	150
c-*fos*	?	2	2.2	Fetal membranes (m, h)	55 (40)[c]
c-*mos*	4	8	–	–	–
c-H-*ras*/c-*bas*	7	11	1.4+4.3	Ubiquitously expressed	21
c-K-*ras* (c-*kis*)	?	12	2.0+4.4	Ubiquitously expressed	21
c-*fms*	?	5	3.7	Placenta (m, h)	?
c-*sis*	15	22	4.2	[Tumor cells] (h)	?
c-*raf*	6	3	3.5+5.5	[Fibroblasts] (m)	?
c-*ski*	?	?	6.0+8.0	Cartilage, muscle, skin (ck)	?
N-*myc*[d]	?	2	3.2	[Tumor cells] (h)	?
N-*ras*	?	1	?	?	?

Abbreviations: ck, chicken; m, mouse; h, human; [ ], no comparative analyses performed

[a] Table complemented 1.3.84

[b] Size estimated from SDS-PAGE

[c] Size predicted from nucleotide sequence analysis

[d] Isolated from human tumor DNA (Kohl et al. 1984)

cation of the biochemical properties of c-*onc* gene products and analysis of their biological activities. Recent findings that the c-*sis* gene product may be platelet-derived growth factor (PDGF) (Waterfield et al. 1983; Doolittle et al. 1983) offer an opportunity to study the function of this c-*onc* protein. Similarly, protein kinase activity associated with several c-*onc* gene products should be helpful in delineating their role in normal cellular processes. (b) Introduction of transcriptionally active normal or mutated c-*onc* genes into different cell types in vitro and into developing animals. A promising development is the identification of several c-*onc* homologs in *Drosophila* and yeast. It has recently been demonstrated that c-*src* (*Drosophila*)-specific sequences are expressed only at specific stages of larval development (M. Simon and J.M. Bishop, personal communication). A 21 K protein can be identified in yeast by immunoprecipitation with antisera raised against v-H-*ras* p21 protein (E. SCOLNICK, personal communication). The genomes of *Drosophila* and yeast can be genetically manipulated to study the role of mutated c-*onc* genes.

To understand the mechanisms of neoplastic transformation, it is imperative that we know more about the role of c-*onc* sequences in normal developmental processes. In the coming years, we can expect a surge of knowledge in this area. Certain problems will command higher priority, for instance: (a) The identification by in situ hybridization techniques and immunofluorescence studies of cell types expressing c-*onc* genes at various stages of development. (b) The intracellular localization of c-*onc* proteins. (c) The identification of new c-*onc* genes, either by isolating new oncogenic viruses or by DNA transfection studies. (d) The mechanisms of activation of the oncogenic potential of cellular oncogenes by tissue-specific transcriptional enhancers or other regulatory sequences. (e) Expression of altered c-*onc* genes in mouse embryos or *Drosophila* and identification of mutants in c-*onc* loci in *Drosophila* to analyze the effect of c-*onc* gene products on normal developmental processes. The introduction of altered genes in to yeast by homologous recombination also offers an excellent opportunity to study the role of c-*onc* genes. A cautionary note should be added here that the function of c-*onc* genes in invertebrates and yeast may have evolved quite differently than in the higher eukaryotes. (f) Finally, the effect of chromosomal translocation, amplification, and c-*onc* gene rearrangements on c-*onc* gene expression: Investigation of the function of c-*onc* genes will form a crucial component of the multifaceted attack on understanding the cause of cancer.

Acknowledgements. We thank Drs. B. Adkins, T. Curran, R. Dulbecco, R. Eisenman, and B. Sefton for comments and suggestions. We are grateful to C. Goller for help in the preparation of this manuscript. This work was supported by funds from NIH and ACS.

References

Abelson HT, Rabstein LS (1970) Lymphosarcoma: Virus-induced thymic-independent disease in mice. Cancer Res 30:2208–2212

Abrams HD, Rohrschneider LR, Eisenman RN (1982) Nuclear location of the putative transforming protein of avian myelocytomatosis virus. Cell 29:427–439

Adams J, Gerondakis S, Webb E, Corcoran LM, Cory S (1983) Cellular *myc* gene is altered by chromosome translocation to immunoglobulin locus in murine plasmacytomas and is rearranged similarly in human Burkitt lymphoma. Proc Natl Acad Sci USA 80:1982–1986

Adams JM, Gerondakis S, Webb E, Mitchell J, Bernard O, Cory S (1982) Transcriptionally active DNA region that rearranges frequently in murine lymphoid tumors. Proc Natl Acad Sci USA 79:6966–6970

Alitalo K, Schwab M, Lin CL, Varmus HE, Bishop JM (1983) Homogenously staining chromosomal regions contain amplified copies of an abundantly expressed cellular oncogene (c-*myc*) in malignant neuroendocrine cells from a human colon carcinoma. Proc Natl Acad Sci USA 80:1707–1711

Andersen PR, Devare SG, Tronick SR, Ellis RW, Aaronson SA, Scolnick EM (1981) Generation of BALB/MuSV and Ha-MuSV by type C virus transduction of homologous transforming genes from different species. Cell 26:129–134

Anderson SJ, Furth M, Wolff L, Ruscetti SK, Sherr CJ (1982) Monoclonal antibodies to the transformation-specific glycoprotein encoded by the feline retroviral oncogene v-*fms*. J Virol 44:696–702

Anderson SM, Hayward WS, Neel BG, Hanafusa H (1980) Avian erythroblastosis virus produces two mRNA's. J Virol 36:676–683

Banerji J, Olson L, Schaffner W (1983) A lymphocyte-specific cellular enhancer is located downstream of the joining region in immunoglobulin heavy chain genes. Cell 33:729–744

Barbacid M, Beemon K, Devare SG (1980a) Origin and functional properties of the major gene product of the Snyder-Theilen strain of feline sarcoma virus. Proc Natl Acad Sci USA 77:5158–5162

Barbacid M, Lauver AV, Devare SG (1980b) Biochemical and immunological characterization of polyproteins encoded for by the McDonough, Gardner-Arnstein, and Snyder-Theilen strains of feline sarcoma virus. J Virol 33:196–207

Barbacid M, Breitman ML, Lauver AV, Long LK, Vogt PK (1981) The transformation-specific protein of avian (Fujinami and PRC-II) and feline (Snyder-Theilen and Gardner-Arnstein) sarcoma viruses are immunologically related. Virology 110:411–419

Barker PE, Ruddle FH, Verma IM (1983) Chromosomal assignment of human proto-oncogenes: c-*fos* is located on human chromosome 2. Proc Natl Acad Sci USA (in press)

Barnekow A, Schartt M, Anders F, Bauer H (1982) Identification of a fish protein associated with a kinase activity and related to the Rous sarcoma virus transforming protein. Cancer Res 42:2429–2433

Bishop JM (1981) Retrovirus oncogenes. In: Brown DD (ed) Developmental biology using purified genes. Academic, New York, pp 515–524

Bishop JM (1983) Cellular oncogenes and retroviruses. Ann Rev Biochem 52:301–354

Bishop JM, Varmus HE (1982) Functions and origins of retroviral transforming genes. In: Weiss R et al. (eds) Molecular biology of tumor viruses, 2nd edn. Cold Spring Harbor Laboratory, Cold Spring Harbor, New York, pp 999–1108

Blair DG, Oskarsson TG, Wood WL, McClements PJ, Fischinger G, Vande Woude G (1981) Activation of the transforming potential of a normal cell sequence: a molecular model for oncogenesis. Science 212:941–943

Boyle WJ, Kipsick JS, Reddy EP, Baluda MA (1983) Identification of the leukemogenic protein of avian myeloblastosis virus and of its normal cellular homologue. Proc Natl Acad Sci USA 80:2834–2838

Brugge JS, Erikson RL (1977) Identification of a transformation-specific antigen induced by an avian sarcoma virus. Nature 269:346–348

Brugge JS, Collett MS, Siddiqui A, Marczynska M, Deinhardt F, Erikson RL (1979) Detection of the viral sarcoma gene products in cells infected with various strains of avian sarcoma virus and of a related protein in uninfected chicken cells. J Virol 29:1196–1203

Bunte T, Greiser-Wilke I, Donner P, Moelling K (1982) Association of gap-*myc* proteins from avian myelocytomatosis virus wild-type and mutants with chromatin. The EMBO J 8:919–927

Capon DJ, Seeburg PH, McGrath JP, Hayflick JS, Edman U, Levinson AD, Goeddel DV (1983) Activation of ki-ras-2 gene in human colon and lung carcinomas by two different point mutations. Nature 304:507–513

Chang EH, Furth ME, Scolnick EM, Lowy DR (1982a) Tumorigenic transformation of mammalian cells induced by a normal human gene homologous to the oncogene of Harvey murine sarcoma virus. Nature 297:479–483

Chang EH, Gonda MA, Ellis RW, Scolnick EM, Lowy DR (1982b) Human genome contains four genes homologous to transforming genes of Harvey and Kirsten murine sarcoma viruses. Proc Natl Acad Sci USA 79:4848–4852

Chattopadhyay SK, Chang EH, Lander ML, Ellis RW, Scolnick EM, Lowy DR (1982) Amplification and rearrangement of *onc* genes in mammalian species. Nature 296:361–363

Chen ISY, Mak TW, O'Rear JJ, Temin HM (1982) Characterization of reticuloendotheliosis virus strain T DNA and isolation of a novel variant of reticuloendotheliosis virus strain T by molecular cloning. J Virol 40:800–811

Chen ISY, Wilhelmsen KC, Temin HM (1983) Structure and expression of *c-rel,* the cellular homolog to the oncogene of Reticuloendotheliosis virus strain T. J Virol 45:104–113

Chen JH (1980) Expression of endogenous avian myeloblastosis virus information in different chicken cells. J Virol 36:162–170

Chen JH, Hayward WS, Moscovici C (1981) Size and genetic content of virus-specific RNA in myeloblasts transformed by avian myeloblastosis virus (AMV). Virology 110:128–136

Coffin JM, Varmus HE, Bishop FM, Essex M, Hardy WD, Martin GS, Rosenberg NE, Scolnick EM, Weinberg RA, Vogt PK (1981) Proposal for naming host cell-derived inserts in retrovirus genomes. J Virol 40:953–957

Colby W, Chen E, Smith D, Levinson A (1983) Identification and nucleitide sequence of a human locus homologous to v-*myc* oncogene of avian myelocytomatosis virus MC29. Nature 301:722–725

Collett MS, Erikson RL (1978) Protein kinase activity associated with the avian sarcoma virus *src* gene product. Proc Natl Acad Sci USA 75:2021–2024

Collett MS, Brugge JS, Erikson RL (1978) Characterization of a normal avian cell protein related to the avian sarcoma virus transforming gene product. Cell 15:1363–1369

Collett MS, Erikson E, Purchio AF, Brugge JS, Erikson RL (1979) A normal cell protein similar in structure and function to the avian sarcoma virus transforming gene product. Proc Natl Acad Sci USA 76:3159–3163

Collins S, Groudine M (1982) Amplification of endogenous *myc*-related DNA sequences in a human myeloid leukaemia cell line. Nature 298:679–681

Collins SJ, Groudine MT (1983) Rearrangement and amplification of c-*abl* sequences in the human chronic myelogenous leukemia cell line K562. Proc Natl Acad Sci USA 80:4813–4817

Cook WD (1982) Rapid thymomas induced by Abelson murine leukemia virus. Proc Natl Acad Sci USA 79:2917–2921

Cooper JA, Hunter T (1983) Regulation of cell growth and transformation by tyrosin-specific protein kinases: The search for important cellular substrate proteins. In: Vogt, Koprowski (eds) Current Topics in Immunology and Microbiology, Vol. 107 Springer, Berlin Heidelberg New York

Cooper GM (1982) Cellular transforming genes. Science 218:801–806

Cooper GM, Neiman PE (1980) Transforming genes of neoplasms induced by avian lymphoid leukosis viruses. Nature 287:656–659

Cooper GM, Okenquist S, Silverman L (1980) Transforming activity of DNA of chemically transformed and normal cells. Nature 289:418–421

Cotton PC, Brugge JS (1983) Neural tissues express high levels of the cellular src gene product $pp60^{c\text{-}src}$. Mol Cell Biol 3:1157–1162

Courtneidge SA, Levinson AD, Bishop JM (1980) The protein encoded by the transforming gene of avian sarcoma virus ($pp60^{src}$) and a homologous protein in normal cells ($pp60^{proto\text{-}src}$) are associated with the plasma membrane. Proc Natl Acad Sci USA 77:3783–3787

Crews S, Barth R, Hood L, Prehn J, Calame K (1982) Mouse c-*myc* oncogne is located on chromosome 15 and translocated to chromosome 12 in plasmacytomas. Science 218:1319–1321

Curran T, Teich NM (1982) Candidate product of the FBJ murine osteosarcoma virus oncogene: characterization of a 55,000 dalton phosphoprotein. J Virol 42:114–122

Curran T, Peters G, Van Beveren C, Teich NM, Verma IM (1982) The FBJ murine osteosarcoma virus: identification and molecular cloning of biologically active proviral DNA. J Virol 44:674–682

Curran T, Miller AD, Zokas FL, Verma IM (1984) Viral and cellular fos gene products are located in the nucleus. Cell 36:259–268

Dalla-Favera R, Gelmann EP, Gallo RC, Wong-Staal F (1981) A human *onc* gene homologous to the transforming gene (v-*sis*) of simian sarcoma virus. Nature 292:31–35

Dalla-Favera R, Wong-Staal F, Gallo RC (1982a) *onc* gene amplification in promyelocytic leukaemia cell line HL-60 and primary leukaemic cells of the same patient. Nature 299:61–63

Dalla-Favera R, Bregni M, Erikson J, Patterson D, Gallo RC, Croce CM (1982b) Human c-*myc* *onc* gene is located on the region of chromosome 8 that is translocated in Burkitt lymphoma cells. Proc Natl Acad Sci USA 79:7824–7827

Dalla-Favera R, Franchini G, Martinotti S, Wong-Staal F, Gallo RC, Croce CM (1982c) Chromosomal assignment of the human homologues of feline sarcoma virus and avian myeloblastosis virus *onc* genes. Proc Natl Acad Sci USA 79:4714–4717

Dalla-Favera R, Gallo RC, Gialongo A, Cruce CM (1982d) Chromosomal localization of human homolog (c-*sis*) of the simian sarcoma virus onc gene. Science 218:686–688

DeFeo D, Gonda MA, Young HA, Chang EH, Lowy DR, Scolnick EM, Ellis RW (1981) Analysis of two divergent rat genomic clones homologous to the transforming gene of Harvey murine sarcoma virus. Proc Natl Acad Sci USA 78:3328–3332 de Klein A, van Kessel AG, Grosveld G, Bartram CR, Hagemeijer A, Bootsma D, Spurr NK, Heisterkamp N, Groffen J, Stephenson JR (1982) A cellular oncogene is translocated to the Philadelphia chromosome in chronic myelocytic leukaemia. Nature 300:765–767

DeLarco JE, Todaro GJ (1978) Growth factors from murine sarcoma virus-transformed cells. Proc Natl Acad Sci USA 75:4001–4005

De Martinville B, Giacalone J, Shih C, Weinberg RA, Francke U (1983) Oncogene from human EJ bladder carcinoma is located on the short arm of chromosome 11. Science 219:498–501

Der CJ, Cooper GM (1982) Altered gene products are associated with activation of cellular rask genes in human lung and colon carcinomas. Cell 32:201–208

Der CJ, Krontiris TG, Cooper GM (1982) Transforming genes of human bladder and lung carcinoma cell lines are homologous to the *ras* genes of Harvey and Kirsten sarcoma viruses. Proc Natl Acad Sci USA 79:3637–3640

Devare SG, Reddy EP, Law JD, Robbins KC, Aaronson SA (1983) Nucleotide sequence of simian sarcoma virus genome: demonstration that its acquired cellular sequences encode the transforming gene product of $p28^{sis}$. Proc Natl Acad Sci USA 80:731–735

Dhar R, Ellis RW, Shih TY, Oroszlan S, Shapiro B, Maizel J, Lowy D, Scolnick E (1982) Nucleotide sequence of the p21 transforming protein of Harvey murine sarcoma virus. Science 217:934–937

Donner L, Fedele LA, Garon CF, Anderson SJ, Sherr CJ (1982a) McDonough feline sarcoma virus: characterization of the molecularly cloned provirus and its feline oncogene (v-fms). J Virol 41:489–500

Donner P, Greiser-Wilke I, Moelling K (1982b) Nuclear localization and DNA binding of the transforming gene product of avian myelocytomatosis virus. Nature 296:262–266

Doolittle RF, Hunkapiller MW, Hood LE, Devare SG, Robbins KC, Aaronson SA, Antoniades HN (1983) Simian sarcoma virus *onc* gene, v-*sis*, is derived from gene (or genes) encoding a platelet-derived growth factor. Science 221:275–277

Downward J, Yarden Y, Mayes E, Scrace G, Totty N, Stockwell P, Ullrich A, Schlessinger J, Waterfield MD (1984) Close similarity of epidermal growth factor and v-*erb*-B oncogene protein sequences. Nature 307:521–527

Ellis RW, DeFeo D, Shih TY, Gonda MA, Young HA, Tsuchida N, Lowy DR, Scolnick EM (1981) The p21 *src* genes of Harvey and Kirsten sarcoma viruses originate from diverent members of a family of normal vertebrate genes. Nature 292:506–511

Ellis RW, DeFeo D, Furth ME, Scolnick EM (1982) Mouse cells contain two distinct *ras* gene mRNA species that can be translated into a p21 *onc* protein. Mol Cell Biol 2:1339–1345

Erikson J, Ar-Rushdi A, Drwinga HL, Nowell PC, Croce CM (1983) Transcriptional activation of translocated c-*myc* oncogene in Burkitt lymphoma. Proc Natl Acad Sci USA 80:820–824

Eva A, Robbins KC, Andersen PR, Srinivasan A, Tronick SR, Premkumar R, Ellmore NW, Galen AT, Lautenberger JA, Papas TS, Westin EH, Wong-Staal F, Gallo RC, Aaronson SA (1982) Cellular genes analogous to retoviral *onc* genes are transcribed in human tumour cells. Nature 295:116–119

Farber E, Cameron R (1980) The sequential analysis of cancer development. Adv Cancer Res 31:125–226

Feldman R, Wang E, Hanafusa H (1983) Cytoplasmic localization of the transforming protein of Fujinama sarcoma virus: salt-sensitive association with subcellular components. J Virol 45:782–791

Finkel MP, Biskis BO (1968) Experimental induction of osteosarcomas. Prog Exp Tumor Res 10:72–111

Finkel MP, Bistis BO, Jinkins PB (1966) Virus induction of osteosarcomas in mice. Science 151:698–701

Finkel MP, Reilly CA Jr, Biskis BO (1975) Viral etiology of bone cancer. Front Radiat Ther Oncol 10:28–39

Franchini G, Even J, Sherr CJ, Wong-Staal F (1981) *onc* sequences (v-*fes*) of Synder-Theilen feline sarcoma virus are derived from noncontiguous regions of a cat cellular gene (c-*fes*). Nature 290:154–157

Frankel AE, Fischinger PJ (1976) Nucleotide sequences in mouse DNA and RNA specific for Moloney sarcoma virus. Proc Natl Acad Sci USA 73:3705–3709

Frankel AE, Gilbert JH, Porsiz KG, Scolnick EM, Aaronson SA (1979) Nature and distribution of feline sarcoma virus nucleotide sequences. J Virol 30:821–827

Frykberg L, Palmieri S, Beug H, Graf T, Hayman MJ, Vennstrom B (1983) Transforming capacities of avian erythroblastosis virus mutants deleted in the erbA or erbB oncogenes. Cell 32:227–238

Fung TYK, Lewis WG, Crittenden LB, Kung HJ (1983) Activation of cellular oncogene c-*erb* by LTR insertion: molecular basis for induction of erythroblastosis by avian leukosis virus. Cell 33:357–368

Garber EA, Kruger JG, Hanafusa H, Goldberg AR (1983) Only membrane-associated RSV *src* proteins have amino-terminally bound lipid. Nature 302:161–163

Gardner MB, Arnstein P, Rongey RW, Estes JD, Sarma PS, Rickard CF, Huebner RJ (1970) Experimental transmission of feline fibrosarcomas to cats and dogs. Nature 226:807–809

Gattoni S, Kirschmeier P, Weinstein IB, Escobedo J, Diana D (1982) Cellular Moloney murine sarcoma (c-mos) sequences are hypermethylated and transcriptionally silent in normal and transformed rodent cells. Mol Cell Biol 2:42–51

Gazdar AF, Chopra HC, Sarma PS (1972) Properties of a murine sarcoma virus isolated from a tumor arising in an NZW/NZB F_1 hybrid mouse. I. Isolation and pathology of tumors induced in rodents. Int J Cancer 9:219–233

Gillies SD, Morrison SL, Oi VJ, Tonegawa S (1983) A tissue specific transcription enhancer is located in the major intron of rearranged immunoglobulin heavy chain gene. Cell 33:717–728

Goff SP, Gilboa E, Witte ON, Baltimore D (1980) Structure of the Abelson murine leukemia virus genome and the homologous cellular gene: studies with cloned viral DNA. Cell 22:777–785

Goff SP, D'Eustachio P, Ruddle FH, Baltimore D (1982) Chromosomal assignment of the endogenous protooncogene c-*abl*. Science 218:1317–1319

Gonda TJ, Sheiness DK, Fanshier L, Bishop JM, Moscovici C, Moscovici MG (1981) The genome and the intracellular RNAs of avian myeloblastosis virus. Cell 23:279–290

Gonda TJ, Sheiness DK, Bishop JM (1982a) Transcripts from the cellular homolog of retroviral oncogenes: distribution among chicken tissues. Mol Cell Biol 2:617–624

Gonda MA, Young HA, Elser JE, Rasheed S, Talmadge CB, Nagashima K, Li C-C, Gilden RV (1982b) Molecular cloning, genomic analysis, and biological properties of rat leukemia virus and the *onc* sequences of Rasheed rat sarcoma virus. J Virol 44:520–529

Goubin G, Goldman DS, Luce J, Neiman PE, Cooper GM (1983) Molecular cloning and nucleotide sequence of a transforming gene detected by transfection of chicken B-cell lymphoma DNA. Nature 302:114–119

Goyette M, Petropoulos CJ, Shank PR, Fausto N (1983) Expression of a cellular oncogene during liver regeneration. Science 219:510–512

Graf T, Beug H (1978) Avian leukemia viruses: interaction with their target cells in vivo and in vitro. Biochim Biophys Acta 516:269–299

Graf T, Oker-Blom N, Todorov TG, Beug H (1979) Transforming capacities and defectiveness of avian leukemia viruses OK10 and E26. Virology 94:431–436

Graf T, von Kirchbach A, Beug H (1981) Characterization of the hematopoietic target cells of AEV, MC29 and AMV avian leukemia viruses. Exp Cell Res 131:331–343

Groffen J, Heisterkamp N, Grosveld F, Van de Van W, and Stephenson JR (1982) Isolation of human oncogene sequences (v-fes homolog) from a cosmid library. Science 216:1136–1138

Hall A, Marshall CJ, Spurr NK, Weiss RA (1983) Identification of transforming gene in two human saracoma cell lines as a new member of the *ras* gene family located on chromosome 1. Nature 303:396–400

Hampe A, Laprerotte I, Galibert F, Fidele LA, Sherr CJ (1982) Nucleotide sequences of feline retroviral oncogenes (v-*fes*) provide evidence for a family of tyrosine specific protein kinase genes. Cell 30:775–785

Hann SR, Abrams HD, Rohrschneider LR, Eisenman RN (1983) Proteins encoded by v-*myc* and c-*myc* oncogenes: Identification and localization in acute leukemia virus transformants and bursal lymphoma cell lines. Cell 34:789–798

Hardy WJ (1980) The biology and virology of the feline sarcoma viruses. In: Hardy, WD Jr et al (eds) Feline leukemia virus. Elsevier/North Holland, New York, pp 79–118

Harper ME, Franchini G, Love J, Simon MI, Gallo RC and Wong-Stall F (1983) Chromosomal sublocalization of human c-*myb* and c-*fes* cellular *onc* genes. Nature 304:169–171

Harris LJ, Lang RB, Maren KE (1982) Non-immunoglobulin-associated DNA rearrangements in mouse plasmacytomas. Proc Natl Acad Sci USA 79:4175–4179

Harvey JJ (1964) An unidentified virus which causes the rapid production of tumours in mice. Nature 204:1104–1105

Hayman MJ, Ramsay GM, Savin K, Kitchener G, Graf T, Beug H (1983) Identification and characterization of the avian erythroblastosis virus erbB gene product as a membrane glycoprotein. Cell 32:579–588

Hayward WS, Neel BG, Astrin SM (1981) Activation of a cellular *onc* gene by promoter insertion in ALV-induced lymphoid leukosis. Nature 290:475–480

Heisterkamp N, Groffen J, Stephenson JR, Spurr NK, Goodfellow PN, Solomon E, Carritt B, Bodmer WF (1982) Chromosomal localization of human cellular homologues of two viral oncogenes. Nature 299:747–749

Heisterkamp N, Groffen J, Stephenson JR (1983a) Isolation of v-*fms* and its cellular homolog. Virology 126:248–258

Heisterkamp N, Groffen J, Stephenson JR (1983b) The human v-*abl* cellular homologue. J Mol Appl Genet 2:57–68

Hoffman-Falk H, Einat P, Shilo B-Z, Hoffmann FM (1983) *Drosophila melanogaster* DNA clones homologous to vertebrate oncogenes: evidence for a common ancestor to the *src* and *abl* cellular genes. Cell 32:589–598

Hunter T, Sefton BM (1980) Transforming gene product of Rous sarcoma virus phosphorylates tyrosine. Proc Natl Acad Sci USA 77:1311–1315

Jansen HW, Rückert B, Lurz R, Bister K (1983) Two unrelated cell-derived sequences in the genome of avian leukemia and carcinoma inducing retrovirus MH2. EMBO J 2:1969–1975

Jansson M, Philipson L, Vennstrom B (1983) Isolation and characterization of multiple human genes homologous to the oncogenes of avian erythroblastosis virus. EMBO J 2:561–565

Jolly DJ, Esty AC, Subramani S, Friedmann T, Verma IM (1983) Elements in long terminal repeat of murine retroviruses enhance stable transformation by thymidine kinase gene. Nucleic Acids Res 11:1855–1872

Jones M, Bosselmann RA, Houtin VJF, Berub A, Fan H, Verma IM (1980) Identification and molecular cloning of Moloney mouse sarcoma virus-specific sequences from uninfected mouse cells. Proc Natl Acad Sci USA 77:2651–2655

Kawai S, Yoshida M, Segawa K, Sugiyama H, Ishizaki R, Toyoshima K (1980) Characterization of Y73, an avian sarcoma virus: A unique transforming gene and its product, a phosphopolyprotein with protein kinase activity. Proc Natl Acad Sci USA 77:6199–6203

Kirsten WH, Mayer LA (1967) Morphologic responses to a murine erythroblastosis virus. J Natl Cancer Inst 39:311–335

Kitamura N, Kitamura A, Toyoshima K, Hirayama Y, Yoshida M (1982) Avian sarcoma virus Y73 genome and structural similarity of its transforming gene product to that of Rous sarcoma virus. Nature 297:205–208

Klein G (1981) The role of gene dosage and genetic transpositions in carcinogenesis. Nature 294:313–318

Klein G (1983) Specific chromosomal translocations and the genesis of B-cell-derived tumors in mice and men. Cell 32:311–315

Klempnauer K-H, Gonda TJ, Bishop JM (1982) Nucleotide sequence of the retroviral leukemia gene v-*myb* and its cellular progenior c-*myb*: the architecture of a transduced oncogene. Cell 31:453–463

Klempnauer K-H, Ramsay G, Bishop JM, Moscovici G, Moscovici C, McGrath J, Levinson AD (1983) The product of the retroviral transforming gene v-*myb* is a truncated version of the protein encoded by the cellular oncogene c-*myb*. Cell 33:345–355

Kohl NE, Kanda N, Schreck RR, Bruns G, Latt SA, Gilbert F, Alt FW (1983) Transposition and amplification of oncogene-related sequences in human neuroblastomas. Cell 35:359–367

Krontiris TG, Cooper GM (1981) Transforming activity of human tumor DNAs. Proc Natl Acad Sci USA 78:1181–1184

Land H, Parada LF, Weinberg RA (1983) Tumorigenic conversion of primary embryo fibroblasts requires at least two cooperating oncogenes. Nature 304:596–602

Langbeheim H, Shih TY, Scolnick EM (1980) Identification of a normal vertebrate cell protein related to the p21 src of Harvey murine sarcoma virus. Virology 106:292–300

Lee CK, Chan EW, Reilly CA Jr, Pahnke VA, Rockus G, Finkel MP (1979) In vitro properties of FBR murine osteosarcoma virus (40650). Proc Soc Exp Biol Med 162:214–220

Levinson AD, Courtneidge SA, Bishop JM (1981) Structural and functional domains of the Rous sarcoma virus transforming protein ($pp60^{src}$). Proc Natl Acad Sci USA 78:1624–1628

Levy JA, Hartley JW, Rowe WP, Huebner RJ (1973) Studies of FBJ osteosarcoma virus in tissue culture: I. Biologic characteristics of the "C"-type viruses. J Natl Cancer Inst 51:525–539

Marcu KB, Harris LJ, Stanton LW, Erikson J, Watt R, Croce CM (1983) Transcriptionally active c-*myc* oncogene is contained within NIARD, a DNA sequence associated with chromosome translocations in B-cell neoplasia. Proc Natl Acad Sci USA 80:519–523

Mathey-Prevot B, Hanafusa H, Kawai S (1982) A cellular protein is immunologically crossreactive with and functionally homologous to the Fujinami sarcoma virus transforming protein. Cell 28:897–906

McBride OW, Swan DC, Santos E, Barbacid M, Tronick SR, Aaronson SA (1982) Localization of the normal allele of T24 human bladder carcinoma oncogene to chromosome 11. Nature 300:773–774

McCoy MS, Toole JJ, Cunningham JM, Chang EH, Lowy DR, Weinberg RA (1983) Characterization of a human colon/lung carcinoma oncogene. Nature 302:79–81

McDonough SK, Larsen S, Brodey RS, Stock ND, Hardy WD Jr (1971) A transmissible feline fibrosarcoma of viral origin. Cancer Res 31:953–956

McGrath JP, Capon DJ, Smith DH, Chen EY, Seeburg PH, Goeddel DV, Levinson AD (1983) Structure and organization of the human ki-ras proto-oncogene and related processed pseudogene. Nature 304:501–506

Miller AD, Curran T, Verma IM (1984) C-fos protein can induce cellular transformation: A novel mechanism of activation of a cellular oncogene. Cell 36:51–60

Mitelman F, Levan G (1981) Clustering of aberrations to specific chromosomes in human neoplasma: IV. A survey of 1,871 cases. Hereditas 95:79–139

Moloney JB (1966) A virus-induced rhabdomyosarcoma of mice. Natl Cancer Inst Monogr 22:139–142

Müller R (1983) Differential expression of cellular oncogenes during murine development and in teratocarcinoma cell lines. Cold Spring Harbor Conf Cell Prolif 10:451–468

Müller R, Slamon DJ, Tremblay JM, Cline MJ, Verma IM (1982) Differential expression of cellular oncogenes during pre- and postnatal development of the mouse. Nature 299:640–644

Müller R, Verma IM, Adamson ED (1983a) Expression of c-*onc* genes: c-*fos* transcripts accumulate to high levels during development of mouse placenta, yolk sac and amnion. EMBO J 2:679–684

Müller R, Tremblay JM, Adamson ED, Verma IM (1983b) Tissue and cell type-specific expression of two human c-*onc* genes. Nature 304:454–456

Müller R, Slamon DJ, Adamson ED, Tremblay JM, Müller DM, Cline MJ, Verma IM (1983c) Expression of c-*onc* genes c-ras^{Ki} and c-*fms* during mouse development. Mol Cell Biol 3:1062–1069

Murray MJ, Cunningham JM, Parada LF, Dantry F, Lebowitz P, Weinberg RA (1983) The HL60 transforming sequences: a *ras* oncogene coexisting with altered *myc* gene in hematopoietic tumors. Cell 33:749–757

Mushinski JF, Potter M, Bauer SR, Reddy EP (1983) DNA rearrangement and altered RNA expression of c-*myb* oncogene in mouse plasmacytoid lymphosarcomas. Science 220:795–798

Naharro G, Tronick SR, Rasheed S, Gardner MB, Aaronson SA, Robbins KC (1983) Molecular cloning of integrated Gardner-Rasheed feline sarcoma virus: Genetic structure of its cell-derived sequence differs from that of other tyrosine kinase-coding *onc* genes. J Virol 47:611–619

Naharro G, Robbins KC, Reddy EP (1984) Gene product of v-*fgr onc*: Hybrid protein containing a portion of actin and a tyrosine-specific protein kinase. Science 223:63–66

Neel BG, Jhanwar SC, Chaganti RSK, Hayward WS (1982) Two human c-*onc* genes are located on the long arm of chromosome 8. Proc Natl Acad Sci USA 79:7842–7846

Neuberger MS (1983) Expression and regulation of immunoglobulin heavy chain gene transfected into lymphoid cells. EMBO J 2:1373–1378

Noori-Daloii MR, Swift RA, Kung H-J, Crittenden LB, Witter RL (1981) Specific integration of REV proviruses in avian bursal lymphomas. Nature 294:574–576

Oppermann H, Levinson AD, Varmus HE, Levintow L, Bishop JM (1979) Uninfected vertebrate cells contain a protein that is closely related to the product of the avian sarcoma virus transforming gene (*src*). Proc Natl Acad Sci USA 76:1804–1808

Oskarsson M, McClements W, Blair DG, Maizel JV, Vande Woude F (1980) Properties of a normal mouse cell DNA sequence (*sarc*) homologous to the *src* sequence of Moloney sarcoma virus. Science 207:1222–1224

Papkoff J, Lai MH-T, Hunter T, Verma IM (1981) Analysis of the transforming gene product of Moloney murine sarcoma virus. Cell 27:109–119

Parada LF, Tabin CJ, Shih C, Weinberg RA (1982) Human EJ bladder carcinoma oncogene is homologue of Harvey sarcoma virus *ras* gene. Nature 297:474–478

Parker RC, Varmus HE, Bishop JM (1981) Cellular homolog (c-*src*) of the transforming gene of Rous sarcoma virus: isolation, mapping, and transcriptional analysis of c-*src* and flanking regions. Proc Natl Acad Sci USA 78:5842–5846

Payne GS, Bishop JM, Varmus HE (1982) Multiple arrangements of viral DNA and an activated host oncogene in bursal lymphomas. Nature 295:209–214

Perbal B, Baluda MA (1982) Avian myeloblastosis virus transforming gene is related to unique chicken DNA regions separated by at least one intervening sequence. J Virol 41:250–257

Perry RP (1983) Consequences of *myc* invasion of immunoglobulin loci: facts and speculation. Cell 33:647–649

Peters RL, Rabstein LS, Van Vleck R, Kelloff GJ, Huebner RJ (1974) Naturally occurring sarcoma virus of the BALB/cCr mouse. J Natl Cancer Inst 53:1725–1729

Ponticelli AS, Whitlock CA, Rosenberg N, Witte ON (1982) In vivo tyrosine phosphorylations of the Abelson virus transforming protein are absent in its normal cellular homolog. Cell 29:953–960

Prakash K, McBride OW, Swan DC, Devare SG, Tronick SR, Aaronson SA (1982) Molecular cloning and chromosomal mapping of a human locus related to the transforming gene of Moloney murine sarcoma virus. Proc Natl Acad Sci USA 79:5210–5214

Privalsky ML, Bishop JM (1982) Proteins specified by avian erythroblastosis virus: coding region localization and identification of a previously undetected *erb*-B polypeptide. Proc Natl Acad Sci USA 79:3958–3962

Pulciani S, Santos E, Lauver AV, Long LK, Aaronson SA, Barbacid M (1982) Oncogenes in solid human tumours. Nature 300:539–542

Purchio AF, Erikson E, Brugge JS, Erikson RL (1978) Identification of a polypeptide encoded by the avian sarcoma virus *src* gene. Proc Natl Acad Sci USA 75:1567–1571

Queen C, Baltimore D (1983) Immunoglobulin gene transcription is activated by downstream sequence elements. Cell 33:741–748

Rapp UR, Goldsborough MD, Mark GE, Bonner TI, Groffen J, Reynolds FH Jr, Stephenson JR (1983a) Structure and biological activity of v-*raf*, a unique oncogene transduced by a retrovirus. Proc Natl Acad Sci USA 80:4218–4222

Rapp UR, Reynolds FH Jr, Stephenson JR (1983b) New mammalian transforming retrovirus: Demonstration of a polyprotein gene product. J Virol 45:914–924

Rasheed S, Gardner MB, Huebner RJ (1978) In vitro isolation of stable rat sarcoma viruses. Proc Natl Acad Sci USA 75:2972–2976

Radke K, Beug H, Kornfeld S, Graf T (1982) Transformation of both erythroid and myeloid cells by E26, an avian leukemia virus that contains the *myb* gene. Cell 31:643–653

Rechavi G, Givol D, Canaani E (1982) Activation of a cellular oncogene by DNA rearrangement: possible involvement of an IS-like element. Nature 300:607–611

Reddy EP, Reynolds RK, Santos E, Barbacid M (1982) A point mutation is responsible for the acquisition of transforming properties by the T24 human bladder carcinoma oncogene. Nature 300:149–152

Reynolds FH, Van de Ven WJM, Stephenson JR (1980) Abelson murine leukemia virus transformation-defective mutants with impaired P120-associated protein kinase activity. J Virol 36:374–386

Robbins KC, Devare SG, Reddy EP, Aaronson SA (1982) In vivo identification of the transforming gene product of simian sarcoma virus. Science 218:1131–1133

Rohrschneider LR (1979) Immunofluorescence on avian sarcoma virus-transformed cells: Localization of the *src* gene product. Cell 16:11–24

Rohrschneider LR (1980) Adhesion plaques of Rous sarcoma virus-transformed cells contain the *src* gene product. Proc Natl Acad Sci USA 77:3514–3518

Rohrschneider LR, Gentry LE (1983) Subcellular location of retroviral transforming proteins define multiple mechanism of transformation. Adv Cancer Res

Roussel M, Saule S, Lagrou C, Rommens C, Beug H, Graf T, Stehelin D (1979) Three new types

of viral oncogene of cellular origin specific for haematopoietic cell transformation. Nature 281:452–455
Ruley HE (1983) Adenovirus early region 1A enables viral and cellular transforming genes to transform primary cells in culture. Nature 304:602–606
Sakaguchi AY, Haylor SL, Shows TB, Toole JJ, McCoy M, Weinberg RA (1983) Human c-K-*ras* proto-oncogene on chromosome 12. Science 219:1081–1083
Santos E, Tronick SR, Aaronson SA, Pulciani S, Barbacid M (1982) T24 human bladder carcinoma oncogene is an activated form of the normal human homologue of BALB- and Harvey-MSV transforming genes. Nature 298:343–347
Schwab M, Alitalo K, Varmus HE, Bishop JM, George D (1983) A cellular oncogene (c-Ki-*ras*) is amplified, overexpressed and located within keryotypic abnormalities in mouse adrenocortical tumor cells. Nature 303:497–501
Scolnick EM, Papageorge AG, Shih TY (1979) Guanine nucleotide-binding activity as an assay for src protein of rat-derived murine sarcoma viruses. Proc Natl Acad Sci USA 76:5355–5359
Scolnick EM, Weeks MD, Shih TY, Ruscetti SK, Dexter TM (1981) Markedly elevated levels of an endogenous sarc protein in a hematopoietic precursor cell line. Mol Cell Biol 1:66–74
Sefton BM, Hunter T, Raschke WC (1981a) Evidence that the Abelson virus protein functions in vivo as a protein kinase that phosphorylates tyrosine. Proc Natl Acad Sci USA 78:1552–1556
Sefton BM, Hunter T, Ball EH, Singer SJ (1981b) Vinculin: a cytoskeletal target of the transforming protein of Rous sarcoma virus. Cell 24:165–174
Sefton BM, Trowbridge IS, Cooper JA, Scolnick EM (1982) The transforming proteins of Rous sarcoma virus, Harvey sarcoma virus and Abelson virus contain tightly bound lipid. Cell 31:465–474
Sheiness D, Bishop JM (1979) DNA and RNA from uninfected vertebrate cells contain nucleotide sequences related to the putative transforming gene of avian myelocytomatosis virus. J Virol 31:514–521
Sheiness DK, Hughes SH, Varmus HE, Stubblefied E, Bishop JM (1980) The vertebrate homolog of the putative transforming gene of avian myelocytomatosis virus: characteristics of the DNA locus and its RNA transcript. Virology 105:415–424
Shen-Ong GLC, Keath EJ, Piccoli SP, Cole MD (1982) Novel *myc* oncogene RNA from abortive immunoglobulin gene recombination in mouse plasmacytomas. Cell 31:443–452
Shibuya M, Hanafusa H (1982) Nucleotide sequence of Fujinami sarcoma virus: evolutionary relationship of its transforming protein with transforming genes of other sarcoma viruses. Cell 30:787–795
Shibuya M, Hanafusa T, Hanafusa H, Stephenson JR (1980) Homology exists among the transforming sequences of avian and feline sarcoma virus. Proc Natl Acad Sci USA 77:6536–6540
Shibuya M, Hanafusa H, Balduzzi PC (1982) Cellular sequences related to three new *onc* genes of avian sarcoma virus (*fps, yes* and *ros*) and their expression in normal and transformed cells. J Virol 42:143–152
Shi C, Ben-Zion S, Goldfarb MP, Dannenberg A, Weinberg RA (1979) Passage of phenotypes of chemically transformed cells via transfection of DNA and chromatin. Proc Natl Acad Sci USA 76:5714–5718
Shih TY, Weeks MO, Young HA, Scolnick EM (1979) Identification of a sarcoma virus-encoded phosphoprotein in nonproducer cells transformed by Kirsten and Harvey murine sarcoma virus. Virology 96:64–79
Shih TY, Papageorge AG, Stokes PE, Weeks MO, Scolnick EM (1980) Guanine nucleotide-binding and autophosphorylating activities associated with the $p21^{src}$ protein of Harvex murine sarcoma virus. Nature 287:686–691
Shilo BZ, Weinberg RA (1981) DNA sequences homologous to vertebrate oncogenes are conserved in *Drosophila melanogaster*. Proc Natl Acad Sci USA 78:6789–6792
Shimizu K, Birnbaum D, Ruley MA, Fasano O, Suard Y, Edlund L, Taparpowsky E, Goldfarb M, Wigler M (1983a) Structure of c-K-ras gene of human lung carcinoma cell line calu-1. Nature 304:497–50
Shimizu K, Goldfarb M, Perucho M, Wigler M (1983b) Isolation and preliminary characterization of the transforming gene of a human neuroblastoma cell line. Proc Natl Acad Sci USA 80:383–387
Shimizu K, Goldfarb M, Suard Y, Perucho M, Li Y, Kamata T, Feramisco J, Stavnezer E, Fogh

J, Wigler MH (1983c) Three human transforming genes are related to the viral *ras* oncogenes. Proc Natl Acad Sci USA 80:2112–2116

Snyder SP, Theilen GH (1969) Transmissible feline fibrosarcoma. Nature 221:1074–1075

Souza LM, Strommer JN, Hillyard RL, Komaromy MC, Baluda MA (1980) Cellular sequences are present in the presumptive avian myeloblastosis virus genome. Proc Natl Acad Sci USA 77:5177–5181

Spector DH, Baker B, Varmus HE, Bishop JM (1978a) Characteristics of cellular RNA related to the transforming gene of avian sarcoma virus. Cell 13:381–386

Spector DH, Smith K, Padgett T, McCombe P, Roulland-Dussoix C, Moscovici C, Varmus HE, Bishop JE (1978b) Uninfected avian cells contain RNA related to the transforming gene of avian sarcoma viruses. Cell 13:371–379

Stanton LW, Watt R, Marcu KB (1983) Translocation, breakage and truncated transcripts of c-*myc* oncogene in murine plasmocytomas. Nature 303:401–406

Stavnezer E, Gerhard DS, Binari RC, Balazs I (1981) Generation of transforming viruses in cultures of chicken fibroblasts infected with an avian leukosis virus. J Virol 39:920–934

Stehelin D, Varmus HE, Bishop JM, Vogt PK (1976) DNA related to the transforming gene(s) of avian sarcoma viruses is present in normal avian DNA. Nature 260:170–173

Tabin CJ, Bradley SM, Bargmann I, Weinberg RA, Papageorge AG, Scolnick EM, Dhar R, Lowy DR, Chang EH (1982) Mechanism of activation of a human oncogene. Nature 300:143–149

Takeya T, Hanafusa H (1983) Structure and sequence of the cellular gene homologous to the RSC *src* gene and the mechanism for generating the transforming virus. Cell 32:881–890

Takeya T, Hanafusa H, Junghans RP, Ju G, Skalka AM (1981) Comparison between the viral transforming gene (src) of recovered avian sarcoma virus and its cellular homolog. Mol Cell Biol 1:1024–1037

Taparowsky E, Suard Y, Fasano O, Shimizu K, Goldfarb M, Wigler M (1982) Activation of the T24 bladder carcinoma transforming gene is linked to a single amino acid change. Nature 300:762–765

Taub R, Kirsch I, Morton C, Lenoir C, Swan D, Tronick S, Aaronson S, Leder P (1982) Translocation of the c-*myc* gene into the immunoglobulin heavy chain locus in human Burkitt lymphoma and murine plasmacytoma cells. Proc Natl Acad Sci USA 79:7837–7841

Teich NM, Wyke J, Mak T, Bernstein A, Hardy W (1982) Pathogenesis of retrovirus-induced disease. In: Weiss R et al (ed) RNA tumor viruses, molecular biology of tumor viruses, 2nd ed. Cold Spring Harbor Laboratory, Cold Spring Harbor, New York, pp 785

Theilen GH, Zeigel RF, Twiehaus MJ (1966) Biological studies with RE virus (strain T) that induces reticuloendotheliosis in turkeys, chickens and Japanese quail. J Natl Cancer Inst 37:731–743

Theilen GH, Gould D, Fowler M, Dungworth DI (1971) C-type virus in tumor tissue of a woolly monkey (*Lagothrix* spp) with fibrosarcoma. J Natl Cancer Inst 47:881–889

Twardzik DR, Todaro GJ, Marquardt H, Reynolds FH Jr, Stephenson JR (1982) Transformation induced by Abelson murine leukemia virus involves production of a polypeptide growth factor. Science 216:894–897

Van Beveren C, van Straaten F, Galleshaw JA, Verma IM (1981) Nucleotide sequence of the genome of a murine sarcoma virus. Cell 27:97–108

Van Beveren C, van Straaten F, Curran T, Müller R, Verma IM (1983) Nucleotide sequence analysis of FBJ-MuSV provirus and c-*fos* (mouse) gene reveals that viral and cellular *fos* gene products have different carboxy termini. Cell 32:1241–1255

van Straaten F, Müller R, Curran T, Van Beveren C, Verma IM (1983) Nucleotide sequence of a human c-*onc* gene: deduced amino acid sequence of human c-*fos* protein. Proc Natl Acad Sci USA 80:3183–3187

Vennström B, Bishop JM (1982) Isolation and characterization of chicken DNA homologous to the two putative oncogenes of avian erythroblastosis virus. Cell 28:135–143

Wang JYJ, Baltimore D (1983) Cellular RNA homologous to the Abelson murine leukemia virus transforming gene: Expression and relationship to the viral sequence. Mol Cell Biol 3:773–779

Wang L-H, Feldman R, Shibuya M, Hanafusa H, Notter MFD, Balduzzi PC (1981) Genetic structure, transforming sequences, and gene product of avian sarcoma virus UR1. J Virol 40:258–267

Wang L-H, Hanafusa H, Notter MFD, Balduzzi PC (1982) Genetic structure and transforming sequence of avian sarcoma virus UR2. J Virol 41:833–841

Ward JM, Young DM (1976) Histogenesis and morphology of periosteal sarcomas induced by FBJ virus in NIH Swiss mice. Cancer Res 36:3985–3992

Waterfield MD, Scrace GT, Whittle N, Stroobant P, Johnsson A, Wasteson A, Westermark B, Heldin C-H, Huang JS, Dual TF (1983) Platelet-derived growth factor is structurally related to the putative transforming protein $p28^{sis}$ of simian sarcoma virus. Nature 304:35–39

Watt R, Stanton LW, Marcu KB, Gallo RC, Croces CM, Ravena G (1983) Nucleotide sequence of cloned cDNA of human c-*myc* oncogene. Nature 303:725–728

Westin EH, Gallo RC, Arya SK, Eva A, Souza LM, Baldua MA, Aaronson SA, Wong-Staal F (1982a) Differential expression of the *amv* gene in human hematopoietic cells. Proc Natl Acad Sci USA 79:2194–2198

Westin EH, Wong-Staal F, Gelmann EP, Dalla-Favera R, Papas TS, Lautenberger JA, Eva A, Reddy EP, Tronick SR, Aaronson SA, Gallo RC (1982b) Expression of cellular homologues of retroviral *onc* genes in human hematopoietic cells. Proc Natl Acad Sci USA 79:2490–2494

Willingham MC, Jay G, Pastan I (1979) Localization of the ASV *src* gene product to the plasma membrane of transformed cells by electron microscopic immunocytochemistry. Cell 18:125–134

Williamham MC, Pastan I, Shih TY, Scolnick EM (1980) Localization of the *src* gene product of the Harvey strain of MSV to the plasma membrane of transformed cells by electron microscopic immunocytochemistry. Cell 19:1005–1014

Witte ON, Rosenberg, Paskind M, Sheilds A, Baltimore D (1978) Identification of an Abelson murine leukemia virus-encoded protein present in transformed fibroblast and lymphoid cells. Proc Natl Acad Sci USA 75:2488–2492

Witte ON, Rosenberg N, Baltimore D (1979a) Preparation of syngeneic tumor regressor serum reactive with the unique determinants of the Abelson murine leukemia virus-encoded P120 protein at the cell surface. J Virol 31:776–784

Witte ON, Rosenberg N, Baltimore D (1979b) A normal cell protein cross-reactive to the major Abelson murine leukemia virus gene product. Nature 281:396–398

Witte ON, Dasgupta A, Baltimore D (1980a) Abelson murine leukemia virus protein is phosphorylated in vitro to form phosphotyrosine. Nature 283:826–831

Witte ON, Goff S, Rosenberg N, Baltimore D (1980b) A transformation-defective mutant of Abelson murine leukemia virus lacks protein kinase activity. Proc Natl Acad Sci USA 77:4993–4997

Wong TC, Lai MMC (1981) Avian reticuloendotheliosis virus contains a new class of oncogene of turkey origin. Virology 111:289–293

Yoshida M, Kawai S, Toyoshima K (1980) Uninfected avian cells contain structurally unrelated progenitors of viral sarcoma genes. Nature 287:653–654

Yoshida M, Kawai S, Toyoshima K (1981) Genome structure of avian sarcoma virus Y73 and unique sequence coding for polyprotein p90. J Virol 38:430–437

Young HA, Shih TY, Scolnick EM, Rasheed S, Gardner MB (1979) Different rat-derived transforming retroviruses code for an immunologically related intracellular phosphoprotein. Proc Natl Acad Sci USA 76:3523–3527

Young HA, Rasheed S, Souder R, Benton CV, Henderson LE (1981) Rat sarcoma virus: Further analysis of individual viral isolates and the gene product. J Virol 38:286–293

Yuasa Y, Srivastava SK, Dunn CY, Rhimm JS, Reddy EP, Aaronson SA (1983) Acquisition of transforming properties by alternative point mutations within c-*bas/has* human protooncogene. Nature 303:775–779

Retroviruses 1

Editors: **P.K. Vogt, H. Koprowski**

1983. 16 figures. V, 146 pages
(Current Topics in Microbiology and Immunology, Volume 103)
ISBN 3-540-12167-6

Contents: *U.G. Rovigatti, S.M. Astrin:* Avian Endogenous Viral Genes. – *S.H. Hughes:* Synthesis, Integration and Transcription of the Retroviral Provirus. – *J.C. Neil:* Defective Avian Sarcoma Viruses. – *N.G. Famulari:* Murine Leukemia Viruses with Recombinant env Genes: A Discussion of Their Role in Leukemogenesis. – *M.J. Hayman:* Avian Acute Leukemia Viruses. – *O.N. Witte:* Molecular and Cellular Biology of Abelson Virus Transformation.

Retroviruses 2

Editors: **P.K. Vogt, H. Koprowski**

1984. 26 figures. 180 pages
(Current Topics in Microbiology and Immunology, Volume 107)
ISBN 3-540-12384-9

Contents: *P. Besmer:* Acute Transforming Feline Retroviruses. – *A. Tereba:* Asymmetric Chromosomal Distribution of Endogenous Retrovirus Loci in Chickens and Mice. – *J.G. Krueger, E.A. Garber, A.R. Goldberg:* Subcellular Localization of pp 60^{src} in RSV-Transformed Cells. – *J.A. Cooper, T. Hunter:* Regulation of Cell Growth and Transformation by Tyrosine-Specific Protein Kinases: The Search for Important Cellular Substrate Proteins. – *J.G. Foulkes:* Phosphotyrosyl-Protein Phosphatases.

Springer-Verlag
Berlin
Heidelberg
New York
Tokyo

Mouse Mammary Tumor Virus

Editors: **P.K. Vogt, H. Koprowski**

1983. 12 figures. VII, 103 pages
(Current Topics in Microbiology and Immunology, Volume 106)
ISBN 3-540-12828-X

Contents: *C. Dickson, G. Peters:* Proteins Encoded by Mouse Mammary Tumor Virus. – *V. Traina-Dorge, J. C. Cohen:* Molecular Genetics of Mouse Mammary Tumor Virus. – *R. Michalides, A. van Ooyen, R. Nusse:* Mouse Mammary Tumor Virus Expression and Mammary Tumor Development. – *G. M. Ringold:* Regulation of Mouse Mammary Tumor Virus Gene Expression by Glucocorticoid Hormones.

The four reviews in this volume cover recent developments in research on mouse mammary tumor virus, concentrating on the structure and expression of the viral genome and its unique response to regulation by glucocorticoid hormones.

Springer-Verlag
Berlin
Heidelberg
New York
Tokyo